COLD WEATHER FLYING

Other TAB Books by the author:

No. 2255 *Passing Your Instrument Pilot's Written Exam*
No. 2276 *Study Guide for the Airline Transport Pilot's Written Exam*

COLD WEATHER FLYING

BY JEFF W. GRIFFIN

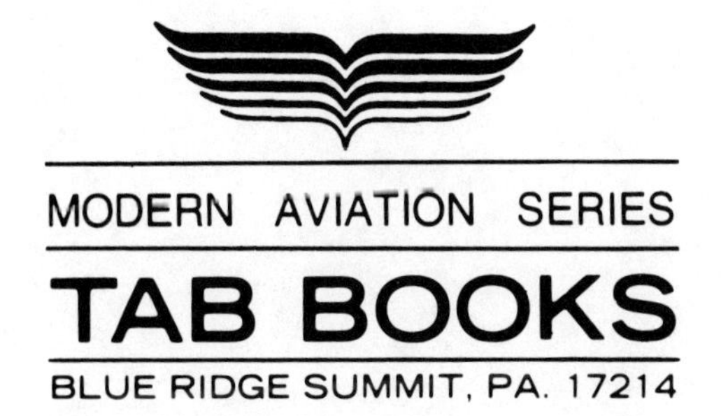

FIRST EDITION

FIRST PRINTING—DECEMBER 1979

Printed in the United States of America

Library of Congress Cataloging in Publication Data

Griffin, Jeff W.
Cold weather flying.

Includes index.
1. Airplanes—Cold weather operation.
I. Title.
TL711.C64G75 629.132'5214 79-26467
ISBN 0-8306-7911-X
ISBN 0-8306-2273-X pbk.

Contents

Acknowledgements

It takes a great deal of learning to become a pilot of several thousand hours. It also takes a great deal of teaching. I'd like to thank the guys that contributed something to my knowledge of cold weather flying. If they should ever see this book they'll know I have not forgotten them, though I have long lost track of them.

Mike Newton
Bob Westmacott
Steve Peters
Rick Nini
Keith Knodt

Introduction

Not too many years ago I made the decision to become a professional pilot. Ever since that decision was made I have taken flying very seriously, even though it is fun. As my knowledge about aviation grew, the more I realized there was to learn. Most of my commercial training was bought in northern Michigan where I lived at the time. Winters are long there. It snows during at least seven months of the year. Cold weather operations are a way of life there.

As a fledgling pilot without much experience, it became clear that gaining experience is a long on-going process. Many times, I thought that to read a book of a practical nature, a head start could be gained. Magazine articles provide lessons. The trouble with magazine articles is their limited scope. At the time I was starting into aviation, I didn't know that in the future I would have the opportunity to write books. I did know, however, what I'd write about if given the chance. All those things that pilots must learn by experience can be put into books. No, it's not the same as the actual doing; but it sure can help one prepare to go out and meet the real world.

When it comes to cold weather, there just aren't too many places that get any colder than northern Michigan. One thing that is so striking about the snow belt is how active the people are during the winter. Even though snow is clogging the roads, the hustle and bustle goes on. The northern folks like snow sports; as long as it is going to snow, we might as well use it for something. This attitude

What are you waiting for? Dust her off and let's go flying!

spills over into their flying habits. I've seen pilots with floatplanes all rigged for summer out doing touch and goes in the snow. On floats! Of course, that is totally acceptable because floats are sturdy. It's just to say that life doesn't stop in the winter. That old plane in the hangar can fly in the cold, thick air on a snowy morning. It just needs a pilot willing to take on the extra challenge of winter. Spark that engine to life some winter and see if cold weather flying isn't a great way to spend time in the winter.

Jeff W. Griffin

Chapter 1
The Case for Winter Flying

As we drove toward the airport, tell-tale signs of the onset of winter were everywhere. The brilliant crimson and gold of autumn leaves were disappearing and the swirl of leaves along the road was like a whirlpool in a bathtub. All the signals were testimony that fall's great flying was going down the drain.

Arriving at the air field this weekend morning ahead of us were friends of the warm weather club. Their hangar doors were pushed aside and all were busy tinkering with their planes. Intrigued by the similarity of their activities, we approached them quietly while observing their work.

"What ya doin' George?"

"Snow in the forecast this afternoon," he shot back, hardly looking up from his chore. "You know it hardly seems like the snow's been gone a month, and it's back again. I can't hardly get any use out of this bird."

"You mean you're going to put her up for the winter?"

"Yep," he motioned toward the sky. "Clouds already tryin' to move in. This late in October just about cinches the end of flying season."

As we walked past the other hangars, the scene was the same. Everyone was interring his aircraft for the cold months.

In the far northern reaches of the country, better known as the snow belt, winter does seem to all but end the normal flying activity of the warm months.

It became clear, after a reasonable amount of thought, that there is some very good flying weather during the cold months. Cold front passage almost always clears the air; and what is better flying than a crisp, clear blue morning with the visibility unlimited? True, the air, as often as not, is a tad on the bumpy side. So what? Aren't there thermals in summer's heat that produce bumps, too?

Another fine time for aviating is after warm front passage. Generally, warm front conditions are the dreariest. Days of lowering clouds, diminishing visibility and various types of precipitation are characteristic. However, as the front passes, the winds turn southwesterly, the clouds rise and usually become scattered to clear. To go flying now would be a good reason to get outside and enjoy the moderating temperatures and sunny hours. It won't last, though—18 to 24 hours usually. Then the onslaught of the next cold front begins.

The beauty of winter days seems to escape the hardened northerner. Most folks start dreaming of the beaches in Florida or the mild desert air of Arizona when the first snow flakes make their debut each year. So, let us consider other reasons to roll the old bird out in the dead of winter.

First, the airplane makes possible quick trips to the sun belt from the northern areas. The best reason in the world to use an airplane is to go somewhere. The trouble is that some cold weather flying is necessary to reach the warmer climates. It becomes obvious, then, that some knowledge of cold weather operations is necessary; and if the trip happens to be a skiing trip to the western states, cold weather flying becomes mixed with mountain flying.

Hidden in the problems of preparing for a cold weather flight is a bonus — vastly improved performance. That extra performance may be a godsend when the wheels are mired in new snow or slush. Mountain air comes down to a more manageable density altitude when winter's air is thick and cold.

In the manuals of most aircraft manufacturers are the performance charts. The charts for calculating takeoff distance, maximum rate of climb and landing distance are all temperature-related. In fact they are predicted on *density altitude* the altitude or "air thickness" at which the aircraft is operating. For instance, an airplane sitting at sea level on a 95° F. day will be operating at an altitude in the neighborhood of 2100 feet. As a result, the takeoff distance will be increased by 13 to 15 percent—a performance penalty. On hot days in the mountains the density altitude may be above the service ceiling of the aircraft you are flying. In that case, takeoff would be ill-advised and more than likely impossible.

Winter, however, turns the tables on that bugaboo. Wintertime is bonus time. Consider a field elevation of 1300 feet and a crisp morning temperature of – 18° F. or – 65° C. Temperatures such as these are realistic in areas such as the Dakotas, the Great Lakes and New England. The density altitude on such a day would be close to 3500 feet below sea level. Since all performance charts consulted were constructed to indicate lack of performance on warm days we can only imply the increase in performance on a cold day. The decrease in takeoff distance on the day described would be close to 25 percent. In other terms, only 75% of the runway that is used on a standard day at sea level would be necessary.

Rate of climb varies according to density altitude, as well. For instance, on a standard day at sea level a Cessna Skylane will climb at 980 feet per minute. That same aircraft will climb at a rate of 1190 feet per minute on that cold morning we described earlier (– 8°F. at 1300 feet elev.). That is an increase in performance of 21 percent. It's enough to make a pilot feel like he's flying a much higher performance airplane. In essence, he is. It's his bonus for braving the freezing temperature.

However, there are some hidden penalties. Colder temperatures aloft will decrease an aircraft's true airspeed. Since the air is thicker at any given altitude, the aircraft is in effect flying below the altitude indicated on the altimeter. This is directly related to the fuel mixture control. If you regularly fly at altitudes of four or five thousand feet, the cold will suck more fuel from your tanks at those altitudes. The strategy to use is to fly higher. Our performance, referring particularly to rate of climb, is increased. The thing to do would be to climb and use that extra climb performance. The air will be smoother, generally. Also, winter winds tend to be at a higher velocity than their summer cousins. So, once again, we have a bonus in reduced block-to-block time when traveling with the wind. That translates into a fuel savings, an important factor in these expensive, energy-conscious days.

When the trip calls for head winds all the way, the logical thing to do is fly lower. Usually, that will mean a compromise between three factors: one, you must consider the velocity of the winds aloft; two, the fuel economy of higher altitudes; and, finally, low level turbulence will play a part in the compromise. The priority which one gives those factors vary; however, turbulence avoidance is probably the most important when carrying passengers. The gusty surface winds after a cold front has just passed, often require an Airmet or even a Sigmet to be issued. On those days, it *is* better judgement to stay on the ground.

Flying out of Cleveland in early March, I learned a lesson about weather briefings and Sigmets. The briefer indicated that a cold front had recently passed and that the route of flight into northern Michigan was clear or peppered with scattered clouds. Included in that briefing were the winds aloft, which were strong and northwesterly. However, everything seemed to indicate going.

We pushed off over Lake Erie and settled into the routine. By the time we had crossed the lake and were in the Detroit area, the bumps were coming harder. They continued rocking the plane until the wheels settled on the pavement. After securing the airplane I called the local FSS and was informed of a Sigmet that was issued two hours prior to our departure time in Cleveland.

I learned two things from that experience. First, that type of weather can be flown, even though the Sigmet sounds ominous. The other thing—always ask if there are any Airmets or Sigmets issued and current for the route of flight.

Punching off in moderate to occasionally severe turbulence is not advisable. For one thing, the plane must be flown at maneuvering speed (V_a) or gust penetration speed (V_b). As a result, cruise performance is wasted and ground speed is low, which only prolongs the discomfort to passengers. Most cold fronts are not that windy after passage. The point is that a pilot can usually handle it alone, but asking passengers to bear it with you is expecting a great deal. Passenger comfort leads my considerations for flight operations.

The other factors mentioned earlier were winds aloft and fuel economy. Head winds cannot always be escaped. It is not likely, with the first 24 hours after passage, that a northbound trip will find tail winds at any altitude. It has been noticed that about 36 hours or so after cold front passage the higher level winds tend to swing around from the south. This is the prelude to the next approaching low pressure center. Flights at 9,000 feet or above may yield a great enough tail wind component to justify the climb; and, once again, with wintertime performance, the climb should be quicker. The fuel economy comes with the territory up there. The key is to know the weather pattern or have the big picture fixed in one's mind. In such a situation, all the important factors are taken care of because the block-to-block time is lower, the air is generally smoother and the fuel burn is lower. That translates into "happy pilot and equally happy passengers." After all, that is what flying is all about.

Utility and the Dollar

There is no doubt that flying is expensive. For that reason most pilots consider the cost of flying before ever purchasing an aircraft.

Fig. 1-1. Now there is a plane ready for some winter camping trips. Check the belly pod for cargo and gear. Its taildragger configuration and large tires will make it easy to handle in the snow.

Many times, dealers and manufacturers, as well as aviation journalists, illustrate the cost as being super-affordable when spread across a gross number of flying hours per year—300 hours a year, for instance. The fact is, very few owners of private aircraft reach that total in a year of flying. Professional pilots fly an average of 750 hours per year, but they are flying several times weekly. So you see, most private pilots don't have that much time to devote to flying. As a result, their hourly cost is higher.

Let's look at the cost of flying a Cessna 152 at today's prices. If you fly a larger plane your costs will be, of course, larger.

The fixed costs of flying are costs one can count on remaining the same. We'll examine those first.

All aircraft should be hangared. That is a matter of opinion, of course, and generally depends on the particular plane. However, when an aircraft is hangared, the upholstery does not fade quickly, the windshield does not easily craze and the exterior paint remains brighter for much longer. These are all cosmetics, true. But, a sharp looking plane sells much quicker, regardless of what the magazine writers say—and they usually say it doesn't matter.

On the other hand, a hangar will aid in preventing corrosion or rusting of various parts that are mechanical. A prime target for rust is the rotor on disc brakes. Hinges on control surfaces can also be affected. All of these, not to mention various external engine parts, benefit from being hangared.

Hangar rents aren't cheap. A local grass airport charges $25 per month for a T-hangar. That sounds like a bargain except for the fact there are no doors on either the north or south side. That won't do a great deal for an airplane during extreme winter conditions. A more realistic price is in the neighborhood of $45 to $55 for a spot in a large hangar. Some T-hangars will run in the same price range. As in simple economics, the cost of hangar rent is directly proportional to how quickly you need one. In the large cities, such as Chicago, New York etc., long waiting lists are habitual.

Insurance is another fixed cost—at least as fixed as any commodity in our inflation-ridden economy. My friend, who briefed me on the cost of ownership is currently (1979) paying $580 per year on one of his Cessna 152's. If you own an older aircraft, some of this cost can be trimmed. Provided the airplane is paid off, hull insurance need not be carried; but, of course, it won't be as easy to replace without insurance in the event of destruction.

It seems truer today than ever before that the good Lord giveth and Uncle Sam taketh away. Airplanes are considered like any other vehicle with the exception that there is a Federal tax annually levied against your airplane. Generally, that tax amounts to $50 though it can be higher on heavier planes. Many states also put their hand in your pocket. State taxes are in the area of $20 to $50 annually. So, we can consider that the tax bill on the airplane will be about $75.

One of the best regulations in the FAR's requires aircraft to be annually inspected. That inspection is the reason why used airplanes maintain their values so well, as compared to automobiles. Once again, though, the cost of an annual inspection contributes to the cost of operation. An annual costs the owner about $125 minimum. That rate may be higher in some parts of the country. It should be pointed out that no repairs or replacement parts are included in that figure.

Another figure of utmost importance is the cost for overhaul. The Cessna 152 has a 108 horsepower Lycoming engine. The cost today to overhaul that powerplant is about $3000. The rub is that the time before overhaul (TBO) is 2,000 hours. By the time this engine needs an overhaul, according to the use of the average private pilot, $3000 will sound as good as the nickel cup of coffee. Just the same, that is the only figure we can use, as no one has learned to predict the prices of the future—that is, except that they will be higher.

The $3000 for that overhaul on a 2,000 TBO engine can be broken down into an hourly figure. Easily, it is $1.50 per hour every hour the plane runs.

The other major cost is fuel which is now standing at 87 cents for 100 octane. Already (April 1979), automobile gasoline has passed that figure. In a Cessna 152, burning about 6 gallons per hour, the cost is $5.22. For the purpose of this discussion we will consider the cost of oil negligible.

Now that all the inherent costs of operating an aircraft have been discussed, let's apply them to the case for winter flying. After talking to several pilots it was discovered that they were flying about 120 hours per year. The figure commonly used is about 300 hours, but most people these days are more occupied with jobs and homes and cannot always "takeoff" as often as they would like. In addition, those 120 hours included about a three or four month dormant period during the winter, a period during which the planes were not flown at all.

What we'll do is examine the cost per hour using the 120 hours our sample pilot is flying in a year even though dormant through the winter months.

$420/yr. @ $35/mo.	Hangar rent
580/yr.	Insurance
75/yr.	Taxes
125/yr.	Annual inspection
$1200/yr.	

Our fixed costs equal $1200 per year or $10.00 an hour figured on our 120 hours. The total hourly cost for this Cessna 152 adding fuel costs and maintenance for overhaul is $16.72 per hour.

Now, let's say the pilot/owner of this aircraft has learned the techniques of cold weather flying. As a result, he can achieve an extra 50 hours in those three or four dormant winter months that would bring the total utility of the aircraft to 170 hours per year. The good news is that the operating cost per hour would now be $13.78—a reduction of nearly $3.00 per hour!

Clearly, the more utility a pilot gets out of the aircraft, the more viable a proposition owning becomes. Hence, winter flying can pay nice tangible benefits.

Finally, along the lines of amenities, comfort is a high priority item. In the midst of July heat, usually little relief can be found in the cabin of an airplane. Relatively few light planes are air conditioned; however, most are heated in the winter. The point is that airplanes can be more comfortable in the winter than in summer.

The points favoring the case for winter flying are clear. Now we must examine the chores and techniques necessary to help us reap the benefits of increased performance, comfort and, above all, reduced operating costs (Fig. 1-1).

Chapter 2
Getting Ready for Winter

When the first nippy air braces your face and your thoughts turn to winter, it's time to prepare for cold weather flying. In this chapter we will discuss the things that can be winterized and the items that should be checked before the main brunt of winter weather arrives. The object of winterizing your airplane is to lower the incidence of lousy eventualities and allow preflight inspections to be shorter and to the point.

All through the ages animals have been guided by instinct to prepare for winter. Some beasts eat voraciously, enabling them to store energy in fat cells which, in turn, enables their bodies to live during hibernation. Other animals stash food in easily accessible hiding places to be used when winter blankets the earth.

The same sort of thought must be given to preparing one's aircraft for winter. Unfortunately, man has no instinct that tells him it is time to go winterize his airplane. Moreover, there is no instinct to tell him what to do to his airplane. Man's saving grace is the ability to reason and consume new information and use the new information to his benefit. With his reason, man can invent ingenious commodities that will help. The point of all this is that one should learn all that is possible on a subject such as winterization, then use one's God-given ability to reason and incorporate original devices and methods that apply to one's airplane.

The items that we will discuss will, if followed, result in a thoroughly winterized aircraft, yet, a pinch of originality will go a long

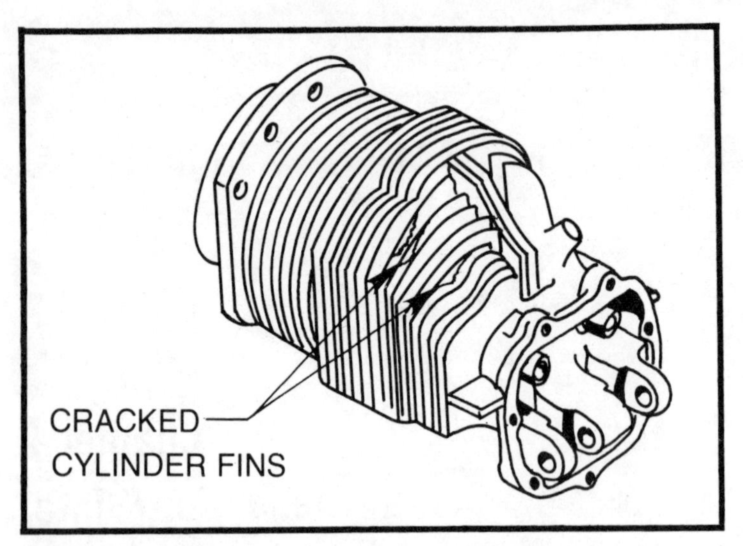

Fig. 2-1. Notice the cracked cylinder fins, they can lead to hot spots.

way. Maintenance costs can be brought down. The ease of movement in winter saves time and that easily translates into money saved.

The Engine Cooling System

Excessive engine heat is always undesirable. Excess heat causes the breakdown of lubricants and, thus, excessive engine wear or damage. The engine cooling system is most important. It is no accident that the engine is cooled by passing air during flight. Cooling systems are carefully engineered for the proper cooling effect.

One gallon of aviation gasoline has a large enough heat value to boil 75 gallons of water. It is easy to see that an engine burning 10 to 20 gallons per hour is producing a great deal of heat. Only about 25 percent of that heat is changed into useful power or transferred to the propeller, the other 75 percent of the manufactured heat must be dissipated in some other manner. In most engines, half of the heat goes out the exhaust and the engine absorbs what's left. Oil circulating through the engine absorbs some of the heat and dissipates it at the oil cooler while the engine cooling system dissipates the rest.

As we said earlier, there is more to cooling than just getting the cylinders in the airstream. A cylinder on the average horizontally opposed engine is approximately the size of a half-gallon milk carton. Outside, the engine metal surface exposed to the airstream is greatly increased by the use of cooling fins. Therefore, it is easier to cool the cylinders. Whenever a cooling fin is broken off, a hot spot within the cylinder is possible. If you find much of one missing on your engine, have a mechanic check it out.

Cowling and baffles are used to direct the air around the cylinder. The baffles are metal and direct the air where it is needed to prevent stagnation of the hot air. Blast tubes are also incorporated to direct jets of cool air at the rear spark plug elbows of each cylinder to prevent the ignition leads from overheating (Fig. 2-1).

An engine can have a problem in the wintertime. That problem is too low an operating temperature. Fuel vaporization and oil distribution and circulation depend directly on a warm engine. In our automobiles, the engine is maintained at the proper operational temperature by a thermostatic valve in the water system. Since not many aircraft engine installations use a water cooling system (though there has been talk of using them again in recent years), other methods have been developed.

A form of temperature control for engines on airplanes is *cowl flaps*. They are usually found on higher performance aircraft. For instance, a Cessna 152, 172, a Cherokee 140, and Warrior don't have these amenities. On the other hand, the Cessna 206, 210 and Piper Seneca do have these devices. Usually, the larger engines are more tightly cowled and require more assistance for cooling. Aircraft

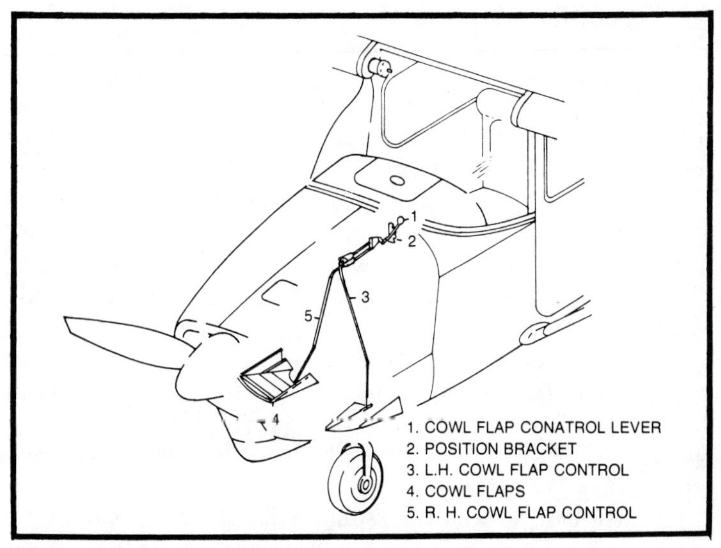

Fig. 2-2. The cowl flaps help keep the engine cool when it is hot and hot when it is cool.

equipped with cowl flaps are great in cold weather mainly because they are more easily kept warm on the ground. The other benefit comes on a low power letdown. They tend to keep the engine warm then, too (Fig. 2-2).

Some of the older aircraft use another cooling method—*augmentor tubes*. Figure 2-3A represents a radial engine; however, aircraft such as the Piper Apache and Aztec use augmentors as well. The augmentors are usually paired and run to the rear of the engine nacelle. An augmentor is actually two tubes, one inside the other. The exhaust collectors feed engine exhaust into the inner augmentor tubes. At this point, the exhaust gas combines with air that has passed around the cylinders. The total mixture is then heated and forms a low pressure, jetlike exhaust. Air that travels through the outer tubes of the augmentors is heated by the inner tubes, but remains uncontaminated by exhaust fumes. This air is used for cabin heating, defrosting and the anti-icing system.

A variation on the augmentor system is the exhaust ejector (see Fig. 2-3B). The design of this system is such that cowl flaps are not needed. The system is able to keep the engine at the proper operating temperature in all regimes of operation.

Now that you know how the cooling systems work in aircraft, what do we do to winterize it? Unfortunately, it isn't as simple as installing more anti-freeze as we do to our cars. Baffles are recommended by some manufacturers to be installed in augmentor tubes. Before one goes to this extreme, do two things. One, be sure the area in which you operate the aircraft is extreme enough in winter to justify the expense and reduction in engine cooling. Second, the FAA requires approval for such installation, so be sure the manufacturer recommends it as well as having FAA approval for the modification.

Winter fronts, which partially cover the cowl nose cap opening and carburetor air intake, are available in winterization kits. Also, non-congealing oil coolers and oil breather insulation can be installed. These items are not recommended unless temperatures frequently range below 20°F.

Whenever these items are installed, a cylinder head temperature gauge is a comforting addition. It isn't mandatory, but it should be. One thing is certain, it will probably keep a pilot from damaging the most expensive part of the plane, the engine, if it is monitored properly.

Cowl flaps need to be checked in the fall. If for some reason they are stuck in one position or another, this could contribute to engine damage. In the heat of the summer it is just as important. If a cowl

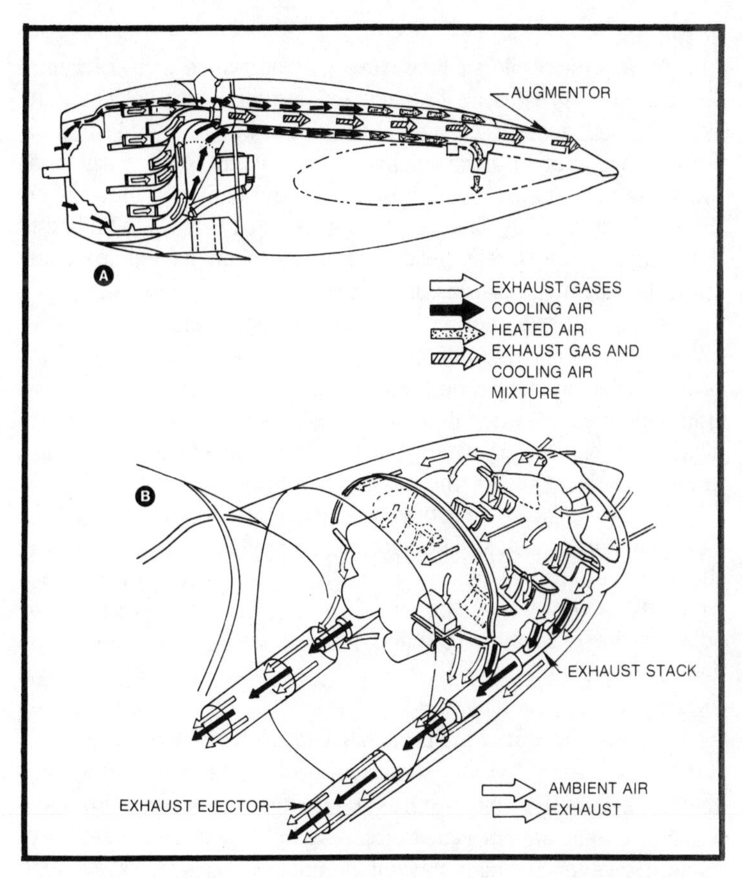

Fig. 2-3. (A) Augmentor tubes are another way to direct cooling air through the engine. (B) A new approach to the old augmentor tubes, exhaust ejectors are found on most modern single-engine aircraft.

flap becomes stuck in the closed position the engine may overheat. In the dead of winter the engine may cool too quickly if the flap is stuck open, because that allows more air circulation through the cowling.

Since cowl flaps are controlled by cables from the cockpit, there are two places they may fail. One is at the hinge in the most forward part of the flap. Another common place is where the cable attaches above the cowl flap, held in place by an eyelet. If the eyelet becomes loose around the cable sheath, the entire cable will move around, instead of the inside wire. It is the inside wire that forces the cowl flap open and shut. The mechanical arrangement is very nearly the same as the throttle arrangement on a lawnmower.

Engine Oil

A great many pilots change their own oil, which is very acceptable. It saves a great deal of money and one can do it as often as one likes. A good idea is to change oil viscosities in winter and summer. Summer weights are always heavier because any liquid or gas tends to become less dense when it is heated and oil is no different. On those scorching 100° days, a pilot wants an oil that can stand up to the temperature. Great peace of mind can be gained from just knowing that those pistons are sliding easily back and forth.

When cold weather approaches, though, that 40W or 50W (W-weight) oil may not do anything to help the engine. Sure, it will provide adequate lubrication when warm, but what about those first traumatic minutes when it is still as thick as wheel bearing grease? That poor old battery, too, has to pull those sticky pistons and crankshaft through. So, the answer is a lower viscosity oil.

Only in the last couple of years have multi-viscosity oils started to appear for airplanes. The automobiling public has had access to them for better than twenty years. The ones that are out, range from 10W to 50W. These oils are a blessing due to the fact that they adjust to the temperature on hand. For instance, a pilot could tank up and fly from New York to Florida and know that his oil is going to do the job in both locations.

Another benefit of multi-vis oils is they don't have to change if the seasons change. For example, a pilot flies mostly in the New England area of the United States and changes his or her oil in middle March. At that time of year there is still plenty of cold weather to come. However, the pilot may not put enough hours on his engine to necessitate an oil change for the summer. If multi-vis oil was used, no oil change would be necessary. Therein, the higher cost of a multi-viscosity oil is offset somewhat by eliminating a change in oils due to seasonal change.

In the days of yore, before multi-vis oils, the oil was thinned in crude fashion. Diesel oil was mixed with regular 40W or 50W oil until it was the right consistency for cold weather use. The problem with that method was the relative inaccuracy of getting the proper consistency and it is not recommended by the author.

Some airplanes in the not so distant past have had oil dilution systems built into the engine. These systems inject aviation gasoline into the oil reservoir where it mixes with the oil and thins it out. The proper viscosity is reached by consulting a chart. For instance, at 0° Farenheit, a thirty-second squirt of gas into the oil sump is needed. The problem with this method of lowering oil viscosity is volume.

Remember, an oil sump has only so much volume available—such as nine quarts. Adding fuel to a full oil sump can cause the oil to overflow whenever the aircraft is rotated to a nose-high attitude. In case your airplane is fitted with this option, pay attention to this warning. The way to eliminate overflowing is to warm the engine thoroughly and give the gasoline in the oil time to evaporate. That will lower the level of the sump sufficiently for normal operation.

One other caution on the use of oil dilution systems: *progessive dilution of the oil is possible.* In order to eliminate this situation, flights of at least two hours duration should be made between dilution operations. The reason is the same as above. This gives the engine a chance to evaporate the avgas and bring the oil back to normal consistency. It should be noted that oil dilution is used to make starting easier only.

Hoses, Clamps and Seals

A very important phase in winter preparation is inspecting various hoses and fittings. The relevancy of this was learned by a friend of mine on a winter day in Michigan. The plane he was flying was a Cessna 172. He bought the aircraft in late spring and had been tying it down outside until a hangar space was available. The run-up in Saginaw was routine and no hint of the incident to come was evident. Having filed IFR for Traverse City in northwestern lower Michigan, he was soon on top droning his way towards a hot supper and a warm fire. Occasional breaks in the clouds revealed the snowy countryside below. As IFR flights go, it was uneventful—a veritable piece of cake.

The trouble began as Minneapolis Center cleared him out of 8,000 feet. The northwestern part of Michigan is famous for its snowfall from lake effect from Lake Michigan. The humidity is generally high and the temperatures are low. Descending through 5,000 feet it was evident that carburetor icing was becoming a threat. "No problem," he thought, "just pull carb heat."

Now, at initial approach altitude, the "172" was well below the clouds. The engine, however, was running poorly and becoming worse. A visual approach was approved and he made a beeline for the airport. Nervously, he yanked on the carburetor heat control once more, with no apparent effect. The engine was shuddering now. On a quarter-mile final, the engine gave its last kick. Just enough for the flare, the plane's altitude and momentum equaled zero and there he sat. He was down on a runway where he was supposed to be and relieved. The airplane? Well, it was towed to a hangar and found to have a worn carburetor heat hose. It had chosen this flight to break into two parts.

The lesson, of course, is that adequate preparation before winter might have revealed the frayed hose.

When I asked him if he had prepped the airplane for winter he said it had never crossed his mind, but he claimed that he'd do one just as soon as the plane was out of the shop.

A pilot has every right to investigate and check his airplane. If you find something wrong with your plane, it is best to check with a mechanic and see if a pilot without an A&P Certificate can repair that item. Another source that is recommended highly is *Lightplane Owner's Maintenance Guide,* TAB book No. 2244.

On the winter prep the hoses should be checked for deterioration. Look for leakage, separation of the cover or braid from the inner tube, cracks, hardening, lack of flexibility and excessive "cold flow." The term "cold flow" refers to those impressions made on hose ends by pressure clamps.

When failure occurs in a flexible hose, one may find it has either swaged end fittings or reusable end fittings. In case the hose has swaged end fittings, the entire assembly must be replaced with factory installed end fittings. The process is simple, or less expensive, when the fittings may be reused. The hose needs only to be replaced with an equally long piece of hose.

Another thing to watch for in checking hoses is the painted stripe. If the stripe spirals around the hose there is unnecessary stress on the nose. Chances are the hose was installed improperly and needs to be replaced, as it may rupture at any time.

Hose clamps and fittings should be checked for tightness and leakage. If you find a metal hose fitting that is loose, don't take any wrench to it. Many times the fittings must be torqued to a manufacturers specification and torque wrench must be used.

Storage Batteries

Batteries are the beginning of the chain of life in an engine. As we stated earlier, if a battery fails on a cold morning, the plane won't be going anywhere very soon. A great deal of anxiety can be removed by proper maintenance and a check in the fall. Batteries that have given some sort of trouble during the warmer months should be replaced. If a battery is more than two years old, it may cause problems as well.

It is hard to replace a battery that is working fine in the fall but is two years old. A new battery may carry the load all winter and a pilot may take it for granted. In fact he may never think about how well the battery is performing. On the other hand, the guy that has to miss an appointment because another battery had to be installed will carry a bad taste in his mouth for some time.

There are two common types of batteries used in aircraft today: lead-acid and nickel cadmium. They are named for the material used in the plates. The batteries have quite different properties and should be maintained according to the manufacturer's specification.

Generally speaking, lead-acid batteries are most widely used in light aircraft. Nickel-cadmium are found most often in aircraft, such as turboprops, that require a prolonged draw of the battery's current to produce a start. Typical turboprop starting cycles are on the order of 25 seconds long.

As could be expected, batteries can overheat as well as freeze. Since we are examining cold weather techniques, let's put overheating aside. Lead-acid batteries exposed to cold temperatures are subject to plate damage when the electrolyte freezes. The electrolyte is the acid-water solution in the battery. The freezing point of the electrolyte is directly related to the specific gravity of the solution, as shown in Table 2-1. To prevent freeze damage, the specific gravity should be maintained at a high level. This can be done by insuring that the proper amount of distilled water is in the battery and that a charge is applied occasionally. Bear in mind that lead-acid batteries are liable to a constant discharge due to the continuous internal chemical action.

The charge can be maintained by an inexpensive charger. Put it in the trickle charge position and the battery will respond on command. Once again, you can see that an electrical outlet in your hangar is extremely useful. Also, occasionally running the engine and charging the battery with the generator are beneficial.

Table 2-1. Lead acid battery electrolyte freezing points.

SPECIFIC GRAVITY	FREEZE POINT °C.	FREEZE POINT °F.
1.300	−70	−95
1.275	−62	80
1.250	−52	−62
1.225	−37	−35
1.200	−26	−16
1.175	−20	− 4
1.150	−15	+ 5
1.125	−10	+13
1.100	− 8	+19

The electrolyte in nickel-cadmium batteries can withstand extreme temperatures. Because that discharge doesn't involve a chemical reaction similar to the lead-acid type. However, the nickel-cadmium electrolyte will freeze at approximately minus 75° F.

How to Check the Battery

Lead-acid batteries are checked with a hydrometer. This instrument measures the specific gravity of the electrolyte in each cell. Batteries manufactured in the United States are fully charged when the specific gravity is between 1.275 and 1.300. A battery that is about one-third discharged will read about 1.240 and one two-thirds discharged will register 1,200 when a hydrometer is used. Before going any further, I ought to clarify what specific gravity is: *It is the weight of the electrolyte as compared to the weight of pure water.*

Checking the battery on a cool fall day could indicate a false reading. To get the true specific gravity, the fluid must be corrected for temperature. Take this example: The electrolyte is 40° F. and the hydrometer reads 1.260. Consulting our chart (see Table 2-2), 16 points must be subtracted to get the proper reading. Thus, the specific gravity is actually 1.244, or the battery is nearing two-thirds discharge.

Also very important when checking a battery is to return the fluid to the *same* cell from which it was collected. The reason is that specific gravities are rarely the same in each cell. Whenever the difference between cells is 0.05 or more, the battery is nearing the end of its useful life.

Nickel-cadmium batteries cannot be checked by using a hydrometer. As we mentioned, the electrolyte's specific gravity does not change with discharge. The only accurate method to determine the state of discharge of a nickel-cadmium battery is by a measured discharge. This takes a special meter.

Fluid levels in a nickel-cadmium battery vary with charge. Hence, a battery of this type must be fully charged and left to stand for at least two hours. After that time the electrolyte may be adjusted by the addition of distilled or demineralized water. Due to the fact that these batteries' fluid levels vary with charge, don't try to fill one while it is installed in the airplane. Spillage or spewage of the electrolyte may occur. The electrolyte is a corrosive and certain parts of the airplane could be "eaten up."

One simple precaution may be more valuable than all other preventive maintenance. If you infrequently fly during the winter,

Table 2-2. Sulfuric acid temperature correction.

ELECTROLYTE TEMPERATURE		POINTS TO BE SUBTRACTED OR ADDED TO SPECIFIC GRAVITY READINGS
°C.	°F.	
60	140	+24
55	130	+20
49	120	+16
43	110	+12
38	100	+ 8
33	90	+ 4
27	80	0
23	70	−4
15	60	−8
10	50	−12
5	40	−16
−2	30	−20
−7	20	−24
−13	10	−28
−18	0	−32
−23	−10	−36
−28	−20	−40
−35	−30	−44

remove the battery and store it at home in a warm place such as heated garage or utility room. Don't forget the battery when you leave for the airport, though.

Cabin Heaters

There are two main types of heaters used in aircraft with the exception of turbine equipment. The most widely used derives its heat from the exhaust of internal combustion engines. The Janitrol type is a cannister which contains combustion within and then distributes it to the cabin.

The aircraft that derive their cabin heat from the engines use shrouds around the muffler or other portions of the exhaust systems. Many of them are tapped at the exhaust ejector tubes, which are shown in Fig. 2-3B. This type can leak carbon monoxide into the cabin, which is the point of focus for winter preparation. Cabin heaters of this type rarely become inoperative; when they do it is usually because the cabin heat control cable has come loose.

The fact that a cabin heater shroud surrounds a muffler means that any perforation in the exhaust system in this area will allow carbon monoxide to be dispersed into the cabin. The exhaust pipe must be exchanged out in order to remedy the problem. Every year several accident investigations name carbon monoxide as a possible or contributing cause to the accident.

Janitrol heaters can leak carbon monoxide, too; however, the main problem is with them becoming dysfunctional. There are several items that can become problems. There are two fans; one fan rams air into the cannister for combustion with fuel while the other fan carries heat away from the outside of the combustion chamber and into the cabin for warmth.

The aviation fuel is pumped into the combustion cannister by a separate fuel pump and ignited by a glow plug. Either device can fail, but it has been my experience that the glow plugs are the trouble makers. A mechanic I checked with said that the emergency heat shutoff switches are most often the culprits.

The point is to check early in the fall and periodically for proper operation. I remember one evening stepping out of the plane on the ramp at Spokane and not feeling my legs from the knees down. A glow plug had gone out en route and that couldn't be prevented. A numbing experience such as that sometimes can be prevented by a good thorough inspection.

One other point should be made: The average pilot should be able to identify the problem with a Janitrol heater. Mechanics recommend not working on them yourself, however, as they might explode leaving you holding a crescent wrench in one hand and no airplane in the other.

Turbine engines supply heat in a totally different manner. They use air that is compressed in the compressor section. As the air is compressed, it is heated and then bled off before it is ever used for combustion—hence, the name "bleed air."

Control Cables

Most aircraft use cables to transfer yoke control inputs to the various flight control surfaces. Because temperature changes from day to day, and day to night, the cables are subject to expansion and contraction and should be adjusted properly by a mechanic, in order that they are able to compensate for the temperature changes encountered. It is a safe bet that if you can hear the cables slapping around in light turbulence, they are too loose. If you take an inspection plate off and pluck one and it really sings, it's too tight.

Other items to inspect are the pulleys. These can become worn and restrict movement or wear out bearings. These can go bad in winter or summer and are a part of annual inspections. They are there to inspect, if you want to take the time.

Oil Pressure Controlled Propellers

Constant speed propellers can give problems in wintertime. Many times I have noticed leaking prop seals on planes that I have

Fig. 2-4. A thorough inspection of the prop seal is recommended before cold temperatures set in and oil congeals.

flown. I am not recommending this practice; the essence of what I am saying is that very small leaks can be tolerated for some time. Those leaks in the winter may be too much for the old seal when the oil gets thick and gooey. A leaky seal should be changed before cold temperatures. The time that a seal would be most likely to let go is during run-up. If a pilot does not allow the oil to warm, the congealed oil may apply more pressure than the seal can stand (Fig. 2-4).

Another time for concern is during training flights on planes with feathering mechanisms. Recurrent practice on multi-engine aircraft cannot be stressed too much. When that training occurs in cold country, the oil may congeal during feathering. Thus, when the engine is restarted, the prop seal goes and so does the engine's oil. Then what was a training exercise, becomes realistic to the max!

A proven aid that has solved this problem is a recirculating oil system for the propeller and feathering mechanism. This system keeps warm oil circulating through the hub of the prop. Whenever the pitch is changed warm air flows in. Since warm oil is thinner than cold or congealed oil, the pressure is proportionately lower which eases the burden on the prop seal.

Wheel Wells and Wheel Pants

Probably the most often encountered problem in winter flying is snow and slush on the runway. Sometimes even rain puddles will cause freezing problems after takeoff. Taxiing and takeoff runs will sling mud and slush into the wheel wells. This may not cause an immediate problem; but, if the temperature is below freezing within the first few thousand feet of climb, the mud or slush could freeze the wheels into place. The problem here is on directional control upon landing (Fig. 2-5).

Imagine a scene like this: On taxiing for takeoff you taxi the aircraft through a low spot with slush in it. As you climb through the first 3,000 feet, the wheel of your fixed gear plane are freezing to the wheel fairing. Upon arriving at your destination there is a stiff 25-knott crosswind blowing and, to make things worse, the paved runway is two-thirds covered with spotty patches of snowpack from the previous plowing.

On final approach you set the airplane up for the crosswind, unaware that your wheels are frozen. The aircraft settles to the runway on a snowpacked area first. A small area of bare pavement lies ahead. As you cross it, you feel a jerk and hear a noise. The aircraft has slowed quite a bit at this point and the rudder has lost most of its effectiveness. The plane begins to weathervane into the wind on the smooth snowpack. You realize something is definitely wrong. There is no directional control at this point. Sliding off the snowpack, in a crab, the tires grab a piece of pavement and the wingtip dips and strikes the ground.

Inspection reveals one wheel fairing is broken badly. That one went when the first bare patch of pavement was encountered. The other wheel is totally intact and still frozen. The wingtip will have to be replaced. Over a thousand dollars worth of repair is needed.

Does this sound like a fantasy? It is not. It is lived several times each winter by unwary pilots. This type of accident could be eliminated by removing the wheel pants for winter operation. If one lives south of the Snow Belt, but travels occasionally to the northland, he or she should be aware of this problem.

Retractable gear airplanes have two different problems. First, the gear can become frozen in the *up* (retracted) position due to slush that was slung into the wheel wells. Second, slush might be splashed on the micro-switches on the gear slide tubes rendering them inoperative. This may indicate one of two possibilities: the gear will not extend or it will extend without a down-and-locked indication on the panel.

If you fly a retractable, there is one thing that can be done to minimize freezing the gear in place. Leave the gear down a little longer than usual and then recycle it. This will allow the wheels to

Fig. 2-5. Eight bolts on this side and one bolt on the other side are all that have to be removed to lift the wheel pant off.

stop spinning normally. In this manner they won't be slinging slush all the way to the wheel well. The recycling of the gear will put the gear in place and then clear it of any obstruction.

Very few problems have ever arisen in my flying career by following these tips on gear operation.

De-ice and Anti-ice Equipment

Airplanes are either certified for flight into known icing conditions, or they are not. The airplanes that are approved are for moderate icing conditions at best. When severe icing conditions are forecast in an area, even the airlines avoid the area.

The airplanes that have no icing equipment, except a heated pitot tube, should not be flown in icing conditions. However, there are methods for doing so if you get caught unexpectedly. We will discuss those in Chapter 5.

Ice equipment is divided into two categories: *De-ice* and *anti-ice.* Anti-ice are those devices a pilot can employ to eliminate problems from the start. For instance, a heated pitot tube is definitely anti-ice. If a pilot waits until ice is encountered, it may be too late. Remember, the pitot tube affects the flight instruments. A pilot that

Fig. 2-6. To check the hot props, just feel the warmth with your hand.

fails to realize something amiss on the panel may fly into the ground before he thinks to flip on the pitot heat.

Some devices, however, are not clearly one or the other. As shown in Fig. 2-6, hot props are situated at the root of the propeller. They are electric coils that warm and melt the ice. Centrifugal force takes care of the rest. If a pilot suspects encountering ice after takeoff, they should be turned on. Ice on the propeller can be a

Fig. 2-7. Hot props are designed to melt the ice and sling it off. Beware the noise as it hits the fuselage.

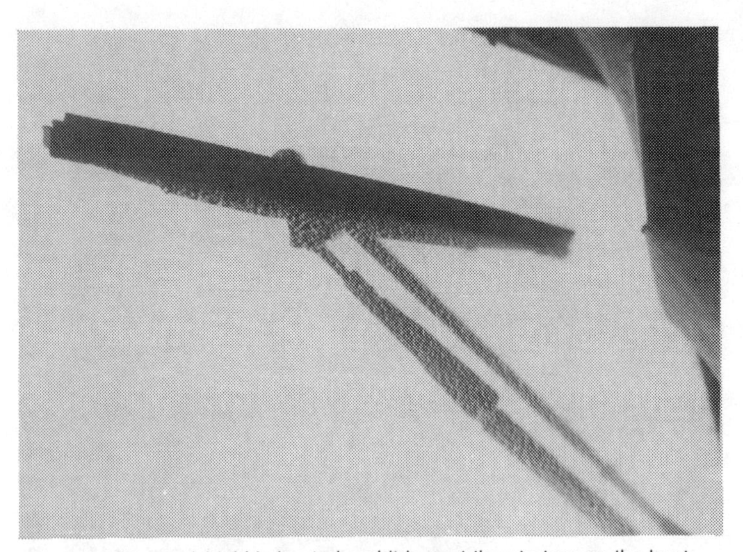

Fig. 2-8. This windshield is heated and it is past time to turn on the heat.

source of many icing problems. If the shape of the airfoil of the propeller becomes distorted, then it will not pull the aircraft along fast enough to maintain flying speed.

Personally, I like to use *heated windshields* as anti-ice equipment because a heated windshield may not be able to clear itself by the landing approach, if I wait too long.

Among the de-ice devices are *inflation boots.* These are found along the leading edges of the wings and horizontal stabilizer. Many times,but not always, they are found on the vertical stabilizer as well. The "boots," as they are called, are inflated from an air accumulator which derives its power from the exhaust or bleed air. They inflate in ripples and the expansion "pops" the ice off.

Among other de-icing equipment are the *alcohol windshield* and *alcohol props*. These are definitely de-icing features. The alcohol sprayed at the roots of props will not prevent ice build-up, but will only aid in removing it. The same goes for the alcohol windshield. When flying an airplane equipped with an alcohol spray windshield, I suggest using the spray intermittently. Don't wait until the approach has commenced and to begin the process of eliminating the ice because alcohol takes time to work. It usually begins by loosening the ice underneath first. Once some ice has melted, the water will mix and form a solution that will work on the rest.

All of these items should be periodically checked for proper operation, as the pilot of the Cessna 310 is doing in Figs. 2-7 through

Fig. 2-9. This Cessna 310 is well equipped with boots on the horizontal and vertical stabilizers.

2-10. Boots are like tires: they get holes in them and will not inflate—in which case they must be repaired.

Pitot heat should be occasionally checked by touching your hand to it. Hot props can be checked in the same fashion, providing the engine isn't operating.

Fig. 2-10. Checking the boots before blasting off into icing conditions is the intelligent thing to do.

On the de-ice alcohol devices, winter preparation is definitely in order. Alcohol spray holes should be cleaned and adjusted for optimum spray patterns and directions. As one might think, all the de-ice and anti-ice should be checked occasionally throughout the winter. The best time is on one of those last, semi-warm fall days, before winter really crunches.

Fire Extinguishers

Fires in an aircraft may happen anytime. The airplane may be on the ground or in the air. Therefore, fire extinguishers need to be kept handy. There should be one in your hangar and one in the aircraft—or several, depending on the aircraft's size. Because fires can happen at anytime, fire extinguishers deserve periodic checks. We will list the items to be checked later, but, first, I'd like to point out the need for fire extinguishers in the airplane. Many planes I've flown, especially in the general aviation fleet, had no extinguisher at all.

As you may know, there are four main classes of fires:

Class A fires are those which occur in ordinary combustible materials such as wood, cloth, paper, upholstery material, etc. Now stop and think for a moment. Does your aircraft have items of these types within its construction? It surely does. A Class A fire, then, is a real possibility.

Class B fires are fires which derive their fuel from flammable petroleum products, greases, solvents and paints. Is there any avgas in your tanks? Then a class B fire is a real possibility, expecially when an engine is flooded or overprimed after it is hot.

Class C fires involve energized electrical components and wiring. Are there any radios, lights or wires on your airplane? Doggone, you could have an electrical fire as well.

Class D fires are those in which metals, such as magnesium, burn. These usually are started by the other classes of fires.

It should be evident that fires could happen in any airplane. If a pilot wants to be prepared he must select a fire extinguisher that will serve the purpose. Fire extinguishers are labeled as to the class of fire for which they are intended. The type I would recommend is a Class A for the passenger compartment and a Class B for the hangar. The Class A fire is the type that would most likely occur in flight. If an electrical fire did occur (Class C) a Class A would work well on items that became inflamed. Usually, a Class C fire is slow-starting and the smell from burning wiring imsulation gives sufficient warning to turn off the electrics and get on the ground (Fig. 2-11).

Be careful when choosing the extinguisher type to be used. For protection of cockpit and passenger compartments, a carbon dioxide or water solution extinguisher is the only type to choose.

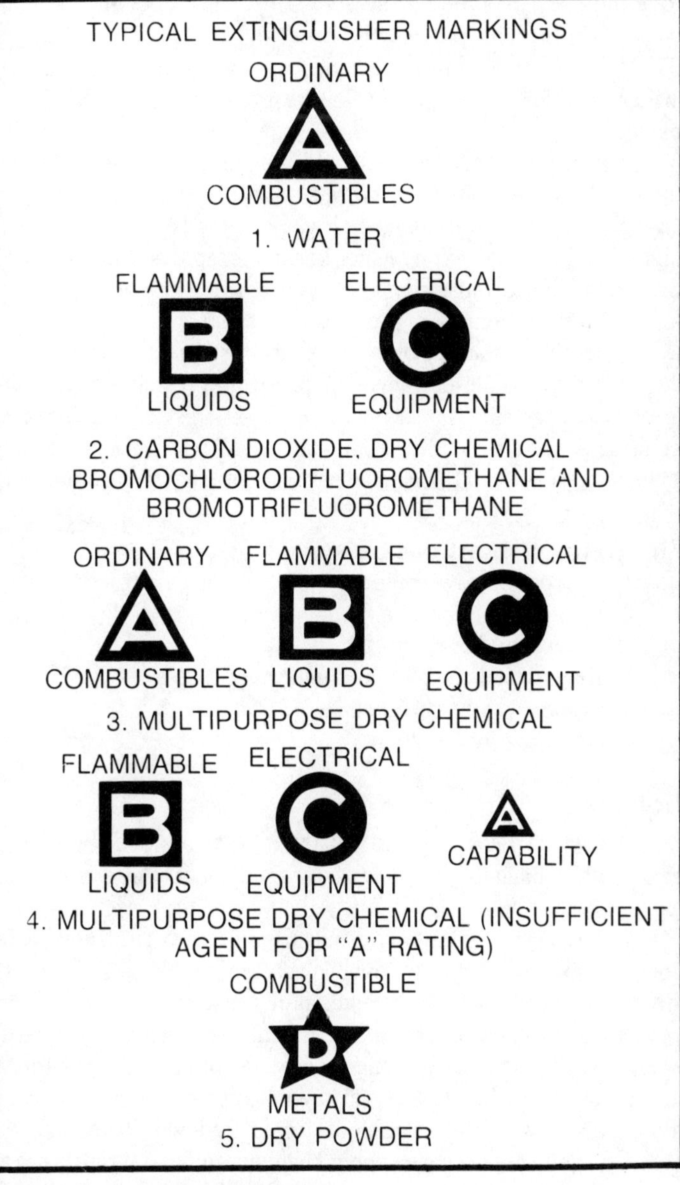

Fig. 2-11. Typical extinguisher markings.

The common aerosol can type of fire extinguisher is definitely not acceptable as an airborne hand-held type extinguisher. In one instance, an aerosol type foam extinguisher, located in the pilot's seat back pocket, exploded and tore the upholstery from the seat, covering the pilot's mother-in-law with patches of cloth and mountains of foam. The pilot never could explain to his wife why he thought that was so funny. Realistically, however, aerosol cans have inadequate volume to fight a fire in proper fashion, and they do explode.

The dry chemical type should not be used either. One was mounted on the wall of an aircraft near a heater vent. For some unknown reason, the extinguisher was reversed end-for-end. As the flight continued, the extinguisher became heated and ruptured. The entire load of passengers was showered with white powder. I don't know if any of them ever rode with us again.

Since the possibility of fire is always with us, we should check the extinguishers periodically. The pre-winter check is a great time to check everything out at once. Below is a list of the necessary items to check on each fire extinguisher. All pilots are comfortable using a checklist and this checklist makes it easy to accomplish the task (Fig. 2-12).

Fire Extinguisher Check List

1. Appropriate extinguisher located in proper place.
2. Safety seals unbroken.
3. All external dirt and rust removed.
4. Gage or indicator in operable range.
5. Proper weight checked.
6. No nozzle obstruction.

Tiedowns

Not everyone can find a hangar if he wants one. Some of us cannot afford hangar fees, and others just don't care whether that old bird is hangared or not.

In any case, we should discuss tying down airplanes. If one doesn't have a hangar, for one reason or another, securing the aircraft for the brisk winter winds is necessary.

Ropes, cables or chains are the material generally used for tiedowns, although cables and chains are often reserved for the heavier airplanes. Most of us will have the occasion to use ropes, so here are a few do's and don'ts of proper tiedown procedure.

On light aircraft, ropes should only be tied to the rings that are provided. The rope should never be tied to a lift strut. The reason for this is, if the rope slips to an area on the strut where there is no slack in the rope, the strut may be bent.

Nylon rope of some form is by far the best type of rope to use. Manila rope will shrink when it becomes wet. If you have to use Manila rope, leave at least an inch of slack. Too much slack, though, will allow the aircraft to jerk against the ropes.

Remember, tight tiedown ropes put inverted flight stresses on the aircraft. These can be considered the same as negative G's.

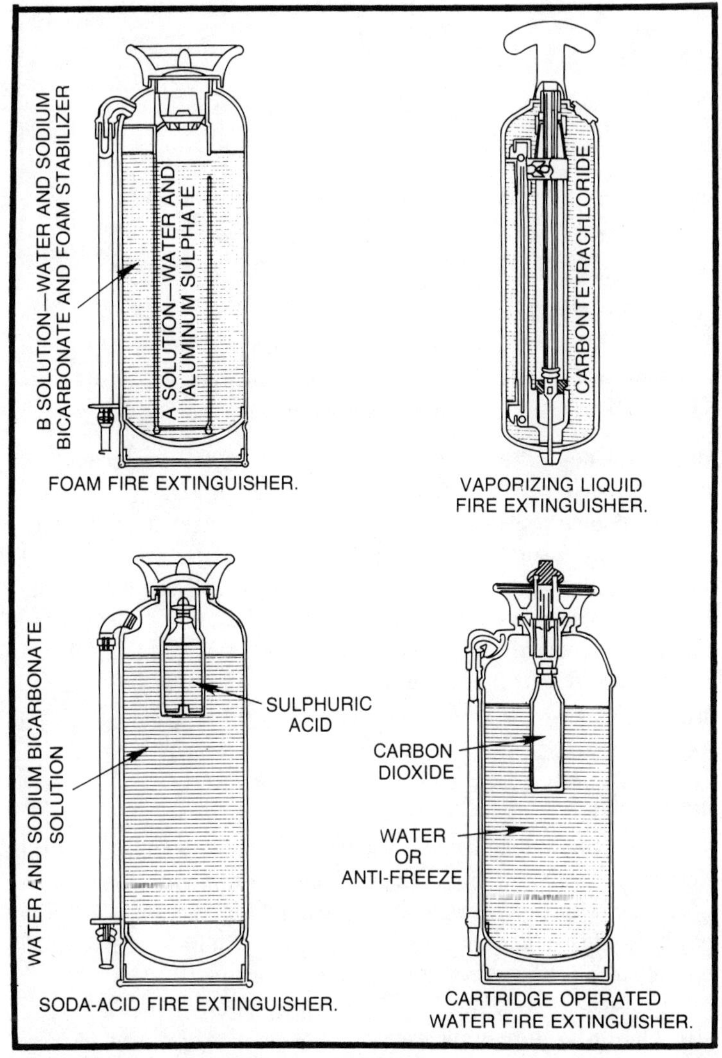

Fig. 2-12. Discontinued types of fire extinguishers.

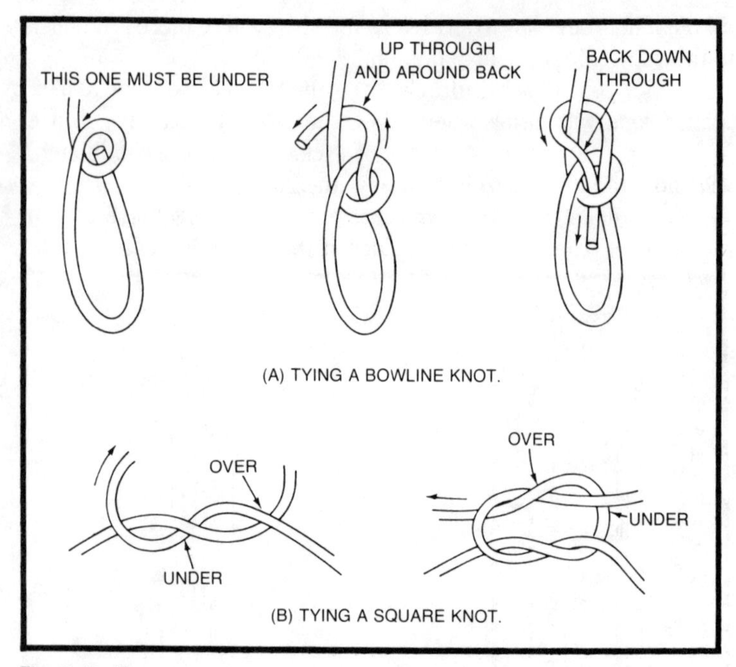

Fig. 2-13. Knots commonly used for aircraft tiedown.

Most airplanes other than aerobatic models are not designed to undergo much stress in this direction.

A tiedown rope will hold the airplane no better than the knot. Anti-slip knots, such as the square knot or bowline (see Fig. 2-13), are quickly tied and are easy to untie. However, if you look around the airport you may see other knots used. I find it somewhat of a hobby just observing the different ways pilots tiedown their planes. Walking down a flight line, it will be rare indeed, if you find two airplanes in a row tied the same way. Take a look sometime and you'll be amused.

Tire Pressures

Some automobile owners will deflate their tires in the winter if they live in the Snow Belt. The reason they will do this is to place more square inches of tread on the ground for traction. While this may be a help in some instances, it is relatively ineffective and it adds excess wear to the tire and lowers gas mileage.

In airplanes, too, it is useless and ineffective. The main reason is that no drive forces whatever are applied to the wheels of an airplane. All locomotion begins with the propeller; hence, all we need is something to roll on until we need to stop. Believe me, a flat, or near flat, tire is going to do nothing for your ability to stop well or with control.

Tires do play an important part in handling the airplane in wintertime conditions. We will explore those techniques in Chapter 4. Meanwhile, mechanics stress using the recommended tire pressures for your plane. Problems can be handled more easily with a good, firm tire.

It may be that you will want to get into winter flying for maximum enjoyment. In this case, skis may be the answer. The skis should be the type approved for your aircraft and should be installed by a certified mechanic. Skis are not usually put on the aircraft until after snow has come to the north country in earnest, which makes fall preparation of skis come late. In order to check your skis, the airplane may need to be put on jacks. If you don't have a way to elevate your plane there are some things that still need to be checked periodically. For instance, some skis are similar to amphibian floats because the tire remains on for use on pavement. The tire protrudes through a slot in the skis; hence, the pilot has the use of pavement, turf or snow-covered landing sights.

Skis are simple installations (see Fig. 2-14). The biggest problem is the shock cord not being tight enough. This will cause a

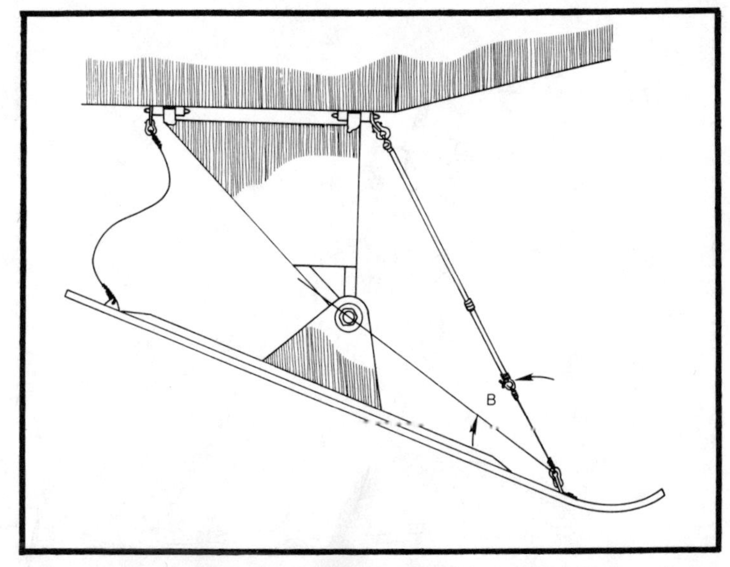

Fig. 2-14. Main ski at a maximum negative incidence. Note the safety cable B is tight; a landing with a skin in this position will be touchy.

negative angle of incidence of the ski. The ski would then be able to dig into the snow, tip first. That might cause the airplane to flip over or to lurch in the direction of the improperly adjusted ski. The best angle of incidence is from zero to five degrees. When that angle is complied with during the pitch up attitude of landing, the ski tip is well out of the snow as the tail makes contact first.

Inspecting a ski installation, one should check the shock cord for wear. Also make sure the safety cable that is strung next to the shock cord is in good shape and allows freedom of movement. The shock cord should be tight, as well. This accomplishes two things: the proper angle of incidence and ski flutter during flight is eliminated.

Whether you fly a tail dragger or a nose dragger the installations on all three ski positions are about the same (as evidenced in Figs. 2-15 and 2-16). The shock cords must be tight on both nose ski and tail ski. The limiting cable on the tail must be loose, as shown.

Putting skis on your snowbound plane adds a new dimension to flying. Many airfields that are closed in the winter are open for a

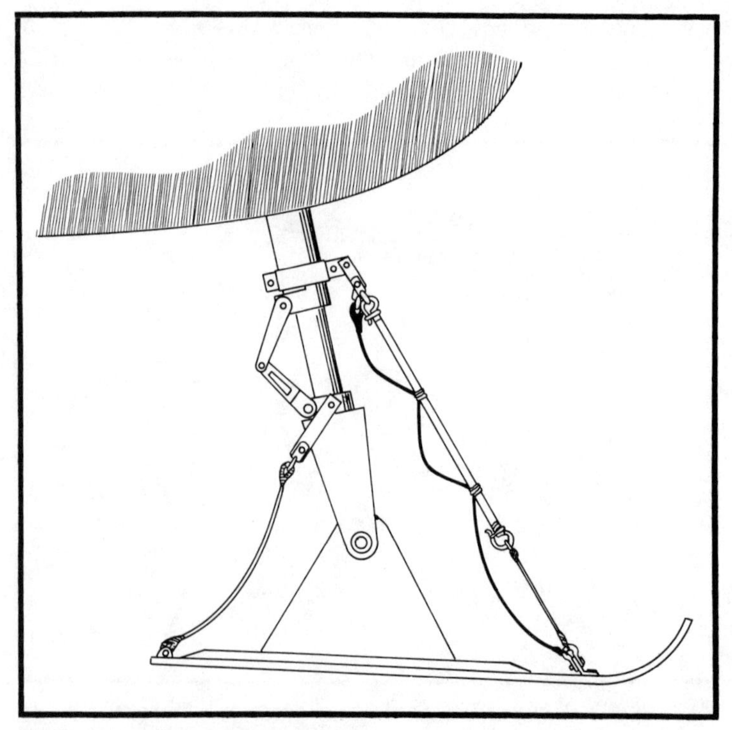

Fig. 2-15. Typical nose ski installation.

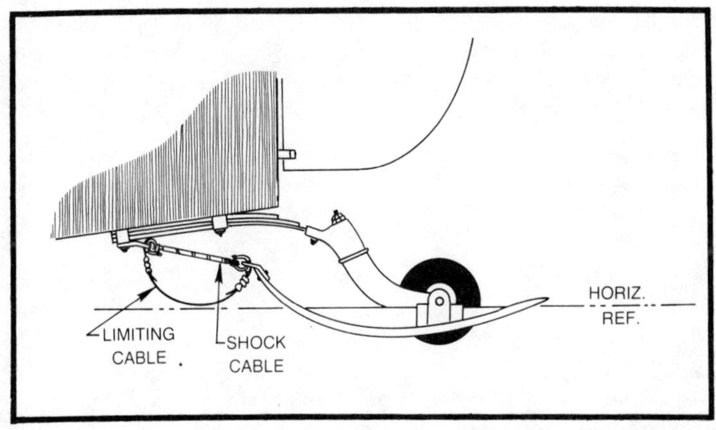

Fig. 2-16. Tail skis are handy because they reduce the wear and tear on the tail wheel assembly.

ski-equipped plane. What could be a better way to get out on a remote lake for some good ice fishing? If you typically use your plane for pleasure trips, maybe you should try skis and feel the spirit of adventure that flying sometimes offers.

Admittedly, some of the things we have discussed can only be discovered by a wary pilot. They cannot be repaired by a pilot unless he has A&P Certificates. The important thing is to take the time in the fall season when the days are still semi-warm and make a thorough inspection. A good idea would be to thumb back through this chapter and jot down the various areas that need to be checked. If you have a list when you get to the airport, the inspection will be quick and to the point.

The old saw, "an ounce of prevention is worth a pound of cure," was never more appropriate. Winterizing an aircraft takes planning and thought. Such thought will make the winter flying season easier to weather and safer. For instance, blasting off into icing conditions is no time to find out the accumulator on the inflation boots has bitten the dust.

As we said in the Chapter 1, winter can be a beautiful time to be aloft. To the pilot who likes a challenge and likes to use his airplane, winter is a good way to whet that appetite. If adequate preparation has been made prior to a flght, cold weather flying is not any more dangerous than flying in July.

Chapter 3

Preflight and Getting Started

If you are like most people, getting out of bed is the hardest part of the day. Below freezing temperatures seem to make the ordeal that much tougher. It seems that leaving that last hot cup of coffee and the warmth of your home, to be slapped in the cheeks with the blistering cold breath of winter, is the second toughest part of the day. However, once out in the elements we begin to adjust to the situation as well as possible. Human beings are quite adaptable and that ability carries us through the rigors of preflight.

Most home-based airplanes are hangared during the winter months. Of course, this is the wisest precaution that can be taken against the winter offensive. As we mentioned earlier, though, hangars are not always available to everyone who wants one. Ordinarily, if you are overnighting in a city in the Snow Belt, hangar space for transients is either at a premium or non-existent. Regardless, of the combination of circumstances, if a pilot flies in the winter, he will eventually be faced with the preflight of an aircraft that has spent the night exposed to nature. The manner in which a pilot goes about conducting the preflight is very important to the safety of the flight.

The ugly reality of eventualities such as the destruction of lives and property is incentive enough to perform preflights that will eliminate the probability of accidents. That incentive, however, is known as a negative incentive. As a flight instructor, I was taught that negative incentives are of dubious value, at best. Let's look for a few positive incentives for conducting the preflight correctly.

When the preflight is done with care and caution we can benefit in ways that every pilot can understand. For instance: if frost on the wings is not removed or polished, we may stall. But when it is removed or polished, we remove the penalty on lift. Thus, we get the performance designed into the wing, reduced drag, which translates into higher true airspeeds and higher fuel economy. Those are some of the positive incentives for handling the preflight walk-around properly.

Preflight Walk-Around

Pulling into the airport parking lot and looking over at your forlorn flying machine standing out in the cold, is enough to make any pilot's heart soar like a rock. Planes just seem to look sad when they stand out during winter. In a few minutes, though, a conscientious pilot can transform that sad and forlorn creature into a vibrant and eager bird, anxious to be airborne.

The first thing to be done at the airplane is to climb into the cockpit. This will get you temporarily out of the wind, if not the cold. Turn the master switch on and check the battery. The reason this should be done first is obvious—you aren't going anywhere if the battery is frozen or discharged. If the battery is in good shape, one of two things can be done. If your aircraft is the type that has electric flaps, run those down to the full down position for the external preflight. This will run current through the battery. Experts recommend this in cold weather because the internal resistance of the battery will cause it to warm slightly and be ready to turn the starter at the given moment. If your airplane has manual flaps, exercise for the battery can still be provided. A check of the rotating beacon and turning the landing light on for about thirty seconds will accomplish the same task. As for those who scoff at this technique and maintain that the voltage drain on the battery from these procedures will only make the battery weak, ask yourselves about summertime preflights. Aren't the lights checked in similar fashion before a flight on a summer evening? They are, of course, unless one is negligent of Federal Aviation Regulations. The fact is, winter *is* harder on batteries; but, a battery will either perform satisfactorily or it won't. This little preflight trick should help.

While you are still sitting in the cockpit, check fuel gauges: and if your airplane is equipped with an auxiliary fuel pump, check its operation. The fuel pressure may be slow to come up into the green arc, due to the temperature. Fuel often takes on a texture approaching that of honey when the temperature is below zero.

Now, it is time to jump out into the open again. It is an unfortunate component of human nature that we do not like uncomfortable temperatures. Therefore, we will dismiss seemingly menial tasks rather quickly to be able to retreat to comfort. The length and thoroughness of preflights are directly proportional to the temperature and wind chill factor. In other words the colder it is the shorter and less thorough they are. It is truly needless to say that attention to detail is more necessary in winter.

Everyone has a method for running a preflight inspection. Some use a checklist. Some folks walk around in a clockwise direction and some reverse that. But an idea that is common to all methods is that they usually start at the pilot's entry point, whether that is left as in Cessnas or right such as in Cherokees.

Generally speaking, the flaps are the first item checked. Since we extended them when we were inside, they are more readily available for complete inspection. As shown in Fig. 3-1 there is snow adhering to the upper surface of the flap. This aircraft was left outside for several days. The snow storm that was in progress as these photographs were taken was the end of a warm spell. The snow that had fallen on this airplane had thawed and then re-frozen as the temperatures dropped. Much of the snow could be brushed away by hand; however, the refrozen material had to be scraped. A good device to carry in your aircraft is a long handled brush and scraper.

Fig. 3-1. The flaps were left down purposely to shed snow. It didn't work and now the flaps must be cleaned before flight.

After the snow is removed satisfactorily, which means *completely*, the hinges and actuating rod should be examined. This is normal procedure in warm weather as well. If this aircraft had been put in a heated hangar to thaw out and then wheeled outside, the checks of the hinges and actuators should be made very carefully. The water will refreeze in these areas and restrict use of these lift-providing surfaces.

The next area to be checked might be the ailerons. Ordinarily, the actuating rod and cotter pins in the hinges are checked. In cold weather this remains true, however, we must be aware that melted snow or ice from the main airfoil will run in the aileron's direction. A slight thaw in the afternoon hours when the sun is shining brightly may cause sufficient melting. The air temperature need not rise above freezing to cause the thaw. The melted water will then run into the hinges and, hence, restrict their movement. It might not be a bad idea to exercise the controls when doing the cockpit inspection. That would break the ice loose or indicate an extreme obstruction to control movement. That way action might be taken before becoming airborne.

How does one free the hinges of ice? The simplest, most economical way is to use deicer fluid. One of the most common brands is I.C.G., ice release coating. That manufacturer claims that its product will aid in removing ice buildups as well as retard the formation of new ice. Windshield de-icer which can be purchased in most service stations, can be very effective, also; but, as with most things associated with aviation, they become more expensive because of that association. Perhaps automobile de-icer could be purchased at a local gasoline station more readily and at a cheaper rate. Commercial deicing fluids such as I.C.G. usually must be ordered from an outlet like Sporty's Pilot Shop on Ohio. If you fly quite a lot in the winter, you could order a case and be ready at any time. If you are like most of us, however, it is nearly impossible to predict how much flying the winter will bring.

Some may try to remove ice from hinges and controls surfaces by using warm water. This is a bad idea, at best, because the water refreezes either on the ground or in the colder temperatures aloft. Can you imagine your flight controls becoming jammed suddenly as you climb out into the colder air? If you can, then don't use warm water to remove ice.

The best method, by far, is to use a hot alcohol spray. Usually a mixture of water and alcohol or methanol is heated by a machine equipped with an air compressor. This mixture should be directed at

Fig. 3-2. A long handled brush is handy for reaching snow way up on the wing. The hand brush is excellent for polishing frost.

leading edges and hinges. The heated fluid is what melts the ice and snow. Alcohol has a very slow effect when not heated. It usually loosens the ice at the adhering surface quickly, but it does not melt the ice per se. The heated fluid melts the ice and then the alcohol mixes with the resulting water to prevent refreezing—that is why this is the best method.

The problem with a heated alcohol spray is availability. Only the airlines and large FBO's tend to have these marvelous contraptions. Many of the smaller outlying airports have no such service and, unfortunately, many of us use the smaller airports because they are more convenient to our destinations. The other problem with a heated alcohol spray is cost. The cost of alcohol and water is very low, but the cost for the service can run from $5 to $40. There is no excuse for the price gouging that goes on. However, they may have you over a barrel. If your bird is iced up and no thaw is expected, the choice becomes one of taking the airlines home or waiting for spring. The best thing to do is pay for the service and then write to an association such as the National Pilots Association or the Aircraft Owners and Pilots Association. These organizations publish lists of

FBO's that frequently provide bad service or charge exorbitant prices for goods and services. In this way, members can steer clear of experiences that leave a bad taste.

Continuing around the airplane on our preflight, we come to the wingtip. We should check the navigation light, as in all preflights, and remove any snow and ice. In newer airplanes, built since the middle 1950's, the wingtips are separate from the main airfoil. At this time, they are generally being manufactured with fiberglas. As a result, the wingtip is not providing any structural integrity in the overall airfoil—that is to say, the wingtips do not support structure loads. The wingtips as we know them today are designed to reduce drag at the end of the airfoil; therefore, snow or ice will create drag that destroys performance.

We are now standing in front of the wing. The leading edge is the "root of all evil" in the winter. Lift begins here and anything interrupting smooth airflow will increase our chances of an accident on takeoff. If the aircraft has recently been in the clouds on an IFR flight, the chances are that there is some ice on the edge. Of course, you'd be aware of this. If the ice isn't too bad, it can be broken off by hand. This really should be done, regardless of the minute amount present.

On low wing airplanes, it is quite easy to see if any snow or ice is on the upper surface of the wing. The snow and ice must be removed, either by scrapping or washing it off with a de-icing agent. In the case of frost the FAA claims that it needs only to be polished. Personally, I feel removal is a better deal. If you choose to polish it though, a medium-soft brush followed by a bath towel will do the job. A high wing aircraft will require more effort and a short ladder. (Fig. 3-2).

About this time on normal preflights we peer into the filler necks of our gas tanks. The wing should be clear of ice before twisting the caps off. Ice and snow can easily be brushed into the tank and that can cause problems. As a matter of fact, that could be a major source of water in the fuel lines. It would be a paradox, indeed, if we thoroughly readied our plane for flight, only to have the engine choke and die from the snow we accidently brushed into the fuel tank (Fig. 3-3).

While we are on the subject of snow removal from the upper surface of the wing, there is one temptation to which we should not fall victim. If one has been so lucky as to have his aircraft hangared in a warm place, falling snow may be a problem. The new falling snow will fall on the wing and if the airplane's skin is warm from being

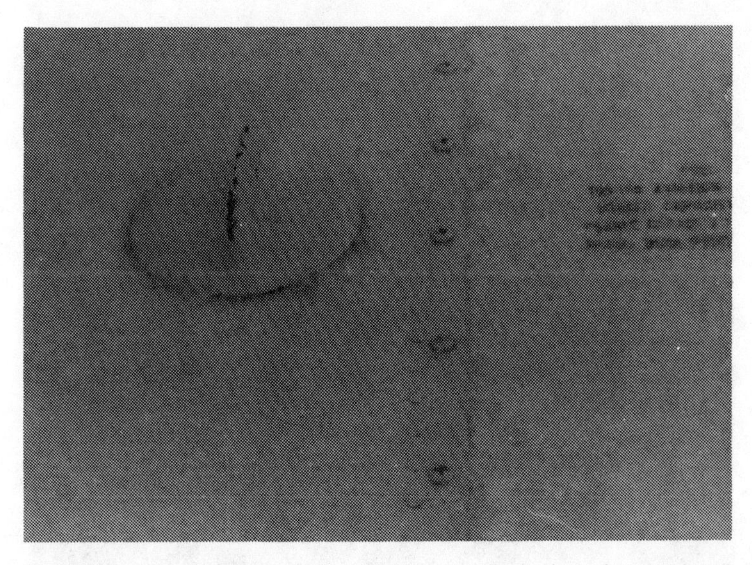

Fig. 3-3. Opening this gas cap before clearing away the ice is a prime way to get ice and thus, water in the fuel.

hangared, some of that snow will melt and then refreeze before the airplane adjusts to the outside air temperature. Although cold snow on a cold surface will blow away, to expect it to do so might be a mistake if the airplane is warm. The top fluff will blow away as takeoff is initiated, but there may be a rime of ice underneath.

There have been times when it was snowing so hard that keeping the wings clean was impossible. One night in Minneapolis during a late winter storm, I had the task of preflighting the Beech Turbine 18 for our cargo run to Chicago. The snow was sloppy, really wet. When I arrived at the plane there were at least two inches of the gunk on the wings. Just before starting, we cleaned it off with a long handled broom. There was no rime ice underneath because the plane was sitting outside all the time. The snow was really coming down so that we could barely see to taxi. By the time we arrived at the threshold of the runway, we had a good inch of snow covering the entire wing again. We had to make a decision: we would roll and not lift off until the snow had blown off the wings. If the takeoff became too long, we would abort. At 90 knots the snow began coming off in chunks and soon the wing was clear, but not for long. The old plane carried one of the heaviest loads of ice I have ever seen into Chicago that night.

There are a couple of things that can be learned from this experience. First, the airplane we were flying had 1100 horsepower

Fig. 3-4. A fuel sample with water in it will look like this. The water will sink to the bottom.

available to help build speed. A small plane with a similar load of snow might not ever have reached the 90 knots we needed to blow the snow away. That is because of lack of power, as well as fresh snow on the runway. The other lesson to be gleaned, is this; if there was any snow left adhering to the wing and then the pilot punched into the icing conditions we did, the odds would be against making a safe landing anywhere.

Another thought occurred to me. We had a low wing airplane. From the start, we could see the snow building on the wings as we taxied. What if a pilot was in a high wing aircraft? First, he would not be as aware of the condition of his aircraft; and second, on takeoff how would he know if the snow was gone? In that case, I'd just go back home or to a motel. It would be too risky.

Let's turn our attention back to our preflight once again. Since we are in the area of the fuel tanks we should check the fuel vents. A fuel vent that is plugged by ice or snow can result in unhappy consequences. The engine may quit just when you need it most after you're airborne or a fuel tank may completely collapse rendering fuel from that tank unavailable for use. Needless to say, all this would result in very expensive damage.

Next to the fuel vents are the fuel tank drains. These are always located in the lowest part of the tank. They are designed this way in order that water or condensation will collect at that point. It is important to visually check the fuel in a clear container. The water will be at the bottom, showing up as little beads or as a glob, depending on the amount present. Although this is a good check, it may not always show the water present. There are experiments on record that have shown that as much as two gallons of fuel had to be drained before any water showed up in the sample bottle. If the plane has set for at least one night, the water should have had sufficient

Fig. 3-5. The ice around the brakes on this retractable should be removed before takeoff. The brakes could heat and melt the ice only to have it refreeze when the wheel goes in the well.

time to make its way to the sump. In conclusion, this test for water may not be perfect, but it's better than a shot in the dark (Fig. 3-4).

As we approach the wing root, we are in the area of the landing gear. As always, we must examine the condition of the tire, the brake rotor (disc) and the brake fluid lines. As shown in Fig. 3-5 ice and snow will accumulate in the areas around the rotor. In this case, the snow is fresh and dry. It would not cause problems and would probably blow away during taxi, however, it would be best to take a long snow brush out. If the area had ice from snow melt or slush from taxi, it will be necessary to warm the area to remove it. The best way is to have the plane towed into a heated hangar and then wait it out. The air from an engine preheater could be directed at the wheel to melt the ice. However, one would have to be convinced that the area was dry before taxiing. There is no point in melting something only to have it refreeze.

On retractable aircraft, the wheel wells should be examined. There is always a possibility that water or slush has splashed into them and become an obstruction. The ice buildup could easily be enough to impede the landing gear from complete retraction. That would cause lack of performance and an unnecessary area for icing in clouds to build. The worst eventually would be for the gear to retract but be bound against the ice so as not to allow the extension of the gear for landing. A short thought given to that malfunction is incentive enough to be thorough with the gear preflight.

One other item should be checked on retractable gear planes. Many have electrically operated gear or sensing systems. The micro switches that close to indicate that the gear is in the down position must be checked clear of ice and snow. A malfunction of these switches will indicate that the gear is not down. That causes skipped heartbeats, huge lumps in the throat, angry stomach and sweaty palms and there's no need for those anxieties (Fig. 3-6).

On hydraulically-activated landing gears, there are usually micro switches, too, generally called "squat switches." They keep the gear from being inadvertently retracted on the ground. If ice should fail these closed, the gear does not retract. The result may not be disastrous, but it is an inconvenience.

We should now be standing in the area of the engine nacelle compartment. Several items need to be checked as usual. The oil level is always checked, of course. In extreme temperatures, the dip stick, once removed, may resemble a popsicle. The oil may be quite congealed and that would mandate preheating. However, it isn't always that cold. On slightly warmer days the oil may move quite

Fig. 3-6. This automatic gear extension sensor is also full of snow. It needs to be cleaned or the gear will hang out all of the time.

easily. It is a little known fact that if the plane is being rocked by the wind, it will cause the oil to creep up the dipstick. Thus, just lifting the stick out and reading it may give you an erroneous idea of the oil level. The dipstick should be wiped off, then plunged back into the crankcase, pulled out again and read for an accurate level.

While you are enjoying the gentle numbing of your feet from the cold, check the fuel sump at the engine. This drain is engineered to be in the perfect position to catch condensation. It is at the lowest point on the fuel line, before the fuel goes into the carburetor or injector. If any contamination is in the line, it should drain here. Once again, have the specimen bottle ready to catch the fuel for examination.

Peering into a semi-dark engine nacelle makes it hard to find problems. If one is not particularly mechanically inclined, the maze of wires and hoses may look like a spaghetti factory. The wires can be loose, for sure, but a person is more likely to see the ones that are charred for some reason. These might need a mechanic's attention.

The hoses, however, are much easier to see and to follow. These usually have three functions: One goes to cabin heat, one to the defroster vents and one to carburetor heat. There are hoses for oil return also. In cold weather, it is not recommended to takeoff when any of these hoses are broken or loose. The carburetor heat is the most important. (We will discuss that later in Chapter 4.) By the same token, if the defroster is disconnected, one might not be able to see outside. Once airborne, that might not be as big of a problem for an instrument rated pilot as a VFR pilot. Finally, if the heater is disconnected, remember the temperatures aloft are usually more extreme.

Before closing the hatch on the engine, we should give the crankcase breather special attention. Frozen breather lines have been attributed to a number of engine failures. When the breather line is frozen shut, pressure builds up in the crankcase. This sometimes results in blowing the filler cap off or rupturing a seal and pumping oil over the side. The combustion of fuel and air results in the by-product of water vapor. Most of this water vapor goes out the exhaust, however, some finds its way to the crankcase. As it cools the vapor condenses in the breather line and cold enough temperatures freeze it. The system can be modified, but it is recommended that an approved change is used to eliminate fire hazard.

On the front of the plane there are several items to be checked. The most important is the propeller. Look for ordinary things, such as cracks and dents. If the propeller is the constant-speed type, streaks of oil on the prop indicate a bad prop seal. The motor mounts can be checked by pulling moderately on the prop. If the engine mounts are broken or have pulled loose, the engine will move inside the nacelle.

The major problem with propellers in the winter is ice. If there is any ice, it should be removed. Actually, the propeller is where all lift begins, so anything that is on the propeller will severely hamper its ability to develop thrust. Ice accumulation on props is rarely symmetrical and that imbalance can cause shaking. Usually, one would find ice on the prop after an IFR flight, where a little ice was encountered on the descent. Most times we would be aware of the ice; but, sometimes, when the flight was the day before, we may have forgotten all about it. The ice might remain if the surface temperatures have remained below freezing since landing. As is human nature, one would be more inclined to jump into the car and head for home or the motel, instead of taking the time to clear the aircraft of ice.

In winter, oil becomes congealed. The best remedy for that situation is heat. However, that may not be possible at all times even though it is recommended. In this case, the prop should be pulled through several times. This sure doesn't warm the oil, but it does unstick the bearings on the crankshaft and piston from the cylinder walls. It is also good for giving a pilot hope that the cold engine will kick off right away.

In the same area of the prop are the air intakes. The air filter for carbureted airplanes (usually a spongy material) collects snow and ice easily. Check it carefully because on carbureted airplanes there is not an alternate source from which the engine can breathe.

Fuel-injected airplanes do have alternate sources for air ingestion. These are normally on the lower side of the nacelle, out of the direct air flow so they should never ice up. If the alternate source is ever used in an emergency, the engine may not develop as much power as usual. That is because the size of the orifice is usually not as large. Also, the effect of ram air is lot and, thus, the manifold pressure will be lower.

The nose gear should give no special problems in winter on fixed gear aircraft. The gear slide tube should be kept clean of mud and properly greased in order that it will move freely. This should be done on the main gear, as well. The retractable nose gear offers a haven for splashed mud and slush. Most nose gears retract towards the rear into a well. Due to the position of the wheel well, behind the wheel, it is an easy target for mud and slush to be slung into it by the tire. As with the other wheels, this well must be checked and cleared of icy obstructions.

On single-engine airplanes, the static port for the flight instruments is often located near the engine. Others are located on the fuselage between the wing and the empennage. For our purposes, let's assume it is forward. The static port has direct control over the vertical speed indicator and altimeter. It is also connected to the airspeed indicator, but only affects its operation indirectly. In flight if the static port is covered or becomes iced over, neither the VSI nor the altimeter will indicate. If one finds the static port plugged or iced over on the preflight, it must be corrected.

Many accidents have occurred because of improper instrument indication; one of the most infamous involved a major airline Boeing 727 climbing out. On this particular flight, the crew forgot to turn the

Fig. 3-7. Always check the pitot tube for heating and obstructions.

pitot heat on. On climb out, the pitot tube iced up and the indication in the cockpit was that the airspeed was increasing. The pilot increased the angle of attack. Still, the airspeed continued to increase. Finally, the aircraft was brought to critical angle of attack. The plane spun uncontrollably from 25,000 feet and the ride was over. Though it was the pitot tube that failed, and not static port in this instance, the message is the same. One of the most critical times in an instrument flight is the transition from visual flight to total instruments. Improperly functioning instruments at this time could easily result in vertigo or misinterpretation. Imagine that and it is evident why the static port is one of the *most* important preflight checks.

On the light and medium twins, the pitot tube is often near, or on, the nose cone. The Cessna 310 pictured has snow inside its pitot tube. Pitot heat will melt it rather quickly and the water should run out the drain hole in the bottom rear. On occasion during the winter months, the pitot heat function should be checked. There is only one sure way to check for proper operation: Turn it on and go outside and feel it (Fig. 3-7).

If the static port or pitot tube is clogged for some reason, clean it out. Instructors always told me to have a mechanic clean it out because he had the proper tools. I adhered to that rule and passed it on to students until it actually happened. The mechanic came over, peered into the dark abyss of the static port and grunted. He soon returned with his special equipment: a paper clip! That cured me of teaching that rule—anybody can clean out a small hole with a paper clip.

On the othe hand, if the static port is clogged or covered with ice, removal isn't so easy. Some sort of heat will have to be applied and the best heat to use is that from a preheater.

Cherokee Arrows have a unique piece of equipment. It is a pitot head for the automatic gear extension system. The only difference is that the holes are much larger. It is positioned behind the pilot's window on the fuselage. The head itself is connected to a diaphragm under the rear passenger seat. When air pressure is applied to the diaphragm, a switch is closed. This causes the gear to remain up or retracted. If anything blocks the orifice to the pitot head, the proper air pressure will not be applied to the diaphragm. As a result, the gear will free fall at all speeds or, rather, will not retract without using the gear override feature. It goes without saying that the ice or snow blocking the gear pitot head will render an inconvenient problem. In summer, as well as winter, this head can become plugged by a smashed bug. I've even heard of one becoming plugged by a piece of gravel being thrown back by the prop.

Fig. 3-8. It isn't easy to see in this photo, but the long wire antenna running to the vertical stabilizer catches a lot of ice on the ground as well as in the air.

Fig. 3-9. This antenna is underneath the fuselage and often goes unnoticed on preflights.

From here we must check the other wing. All things that we looked at on the first side must be duplicated. Once again, those items are fuel quantity and type, gear and micro switches and fuel vents.

Along the cabin and fuselage are various antennas. The old long wire ADF antennas are the worst ice catchers. (Picking up a load of ice going into Boise, Idaho one night, the wire antenna broke loose from the vertical stabilizer. The antenna and insulator flogged the old Aztec all the way in.) The antenna of which I speak usually runs from the front top of the cabin to the top of the vertical stabilizer. Ordinarily, these won't catch snow or ice on the ground. The exception would be in a freezing rain, in which case, the entire airframe is covered and needs to be thawed in the heated hangar (Fig. 3-8).

Other antennas such as marker beacons, transponders and VHF communications can become iced in flight. Due to their relative obscurity (most are underneath) on the fuselage, an icy antenna may go unnoticed in a hasty preflight. The result becomes extra drag, a platform for further quick icing in clouds and the chance of losing the use of those particular radio faculties (Fig. 3-9).

Removing the ice from these antennas is a simple matter. If ice exists on the antenna, just rap it moderately with your hand or knuckle. The ice will crack and flake off easily. If there is a little ice on the wire-type ADF antenna, pluck it as one would a cello string and the ice will fall off.

At the rear of the aircraft is the empennage—in other words, the horizontal and vertical stabilizers. Some of these are flying airfoils, some are not. Some have the standard elevator setup. Even though engineering differences are apparent in this area, some things remain the same from the preflight aspect. There are always control rods, actuating rods and cable attachments to be checked.

Figure 3-10 shows the stabilator area of a Cherokee Arrow. It is readily apparent that the rimy ice must be removed before this airplane is ready for flight. The rime ice on this surface is the type we discussed earlier. The snow was deposited on the surface and melted during warmer days. When the photograph was taken, it was about 20° F. and snowing again. The important item that this photograph illustrates is the snow and ice accumulation on the pegs that stops the stabilator from moving too far in either direction. If this ice isn't somehow removed, the stabilator will not have its full range of travel available to the pilot. The implications of this could be either not enough forward control travel to break a stall or inadequate control travel for a nose-up attitude for landing. Landing on the nose gear first sets up a porpoising motion which is difficult to stop with power alone (Fig. 3-11).

That concludes the walk around inspection. As you can see, there is much more to check in cold weather.The introduction of frozen precipitation into a situation makes a preflight a frustrating situation. One things is for sure: ice and snow that need to be removed prolong our exposure to the elements. Murphy's Law always seems to be in effect in the winter. The one thing to re-

Fig. 3-10. Another surface to clean—note the snow in the hinge area.

member is, if the preflight is cut short or neglected, Murphy's Law has more of a base on which to work.

Preheating and Hangars

There is no substitute for a warm hangar, but you may have to settle for an unheated one. This might not be as bad as it sounds. There are advantages to hangaring, cold or warm. Cold hangars, at least, will keep the snow and ice off the aircraft. If it is snowing when you roll your bird out into the elements, the airplane's skin will be close to the ambient air temperature. In that case, the problem of clearing the flight surfaces of snow will not be as difficult. The snow should brush off easily and leave no rimy ice underneath.

If you find yourself choosing a hangar and the only ones available are unheated, there are some things that will make winter easier. Basically, you should try to find a T-hangar. A large hangar with many airplanes has much traffic and the planes continually get shuffled around. As a result, your plane could develop some unsightly blotches known as "hangar rash." That is why the T-hangar is the best choice. Also, T-hangars often come with a few amenities. Foremost is electricity. An electrical outlet will supply power for preheating devices that are electrically operated. Another is a winch. These are marvelous inventions, especially when the ramp in front is slick with ice or snowpack. There are two things that don't mix: leather-soled shoes and ice. That combination is slicker than a greased pig in a rain storm. A winch pulls the plane back into the hangar easily when one person on ice could not handle the task. If the hangar does not have a winch, one may be installed. If electricity is available, the matter is simple; but if there is not electrical power, a gasoline-powered winch may be an option.

Another desirable amenity is an electric door opener. Many times I have had to chop ice away from sliding door runners. That tires a pilot before he ever gets into the airplane. As a rule, stay away from sliding doors if possible. The best possible type is the kind that folds in half. The pilot that figures out how to have the winch open the door and pull the airplane out and back in will have the aviation community beating a path to his door.

There are a few things that can be used that will eliminate the need for a thorough engine preheating or any preheat at all. One of these devices is automotive in origin. It is a hot dipstick. Many people use them in their cars and trucks during the winter in the Snow Belt. Though these are not designed to fit aircraft engines, most will work well enough. The heating element is about three

Fig. 3-11. The stall warning should be able to move freely and easily.

inches long and is situated at the bottom of the stick. One mechanic said he used that method of keeping the engine oil warm, but he stressed that the entire heating element should be immersed to minimize the chances of an oil fire.

Another method, that isn't very highly recommended, is using a light bulb. Though I have never seen a light bulb explode, mechanics say that if the bulb is placed against metal the possibility exists. Unfortunately, for the best use from a bulb it needs to be placed on top of the cylinder heads or engine casing.

Light bulbs have been used to warm the cockpit area. It is just as important to heat the flight instruments, or at least keep them warm. The problem with a light bulb in the cockpit is that they have been the causes of fires. One fell on a seat and that was all that was needed to burn the plane completely.

An ingenious idea is to use an old electric blanket. These are relatively safe. It won't work in the cockpit, but it will warm an engine. This method works best on airplanes designed with removable top cowlings, such as many Piper products. The secret to making this method work is to be sure and use another blanket to insulate the electric blanket. If this is not done, the electric blanket will never heat. And for the folks that have twins, I guess you need two electric blankets.

The idea of using electric heaters in the cockpit is neither a new nor a good one. There have been many accidents that were caused by an electric space heater overheating a certain area. This can happen easily, if the heater is left unattended. A heater can topple rather easily, too, and that is when most of the problems occur.

There are some other methods for keeping the engine warm throughout the night. These fall into the more expensive category but become a small investment for the pilot serious about winter flying. One system utilizes heating elements incorporated in the cylinder head temperature probes and the oil screen. This method is no good for a fast preheat; but if you are lucky enough to have electricity in your hangar, the 110-volt outlet will supply the power to have your airplane ready at daybreak. The heating elements are permanently installed and, therefore, will be available wherever one travels—helping to make arriving at an outback airport a little less of an ordeal. At least, all one needs is an electrical outlet to keep the airplane eager. This marvelous device is available from the good people at Tanis Aircraft Services, P.O. Box 117, Glenwood, Minnesota 56334. The price ranges from $164 to $293 and does not include installation.

Along the same line of keeping the engine warm through the night is the "Hot Blast" engine warmer (trade name) which is also electric. It is horseshoe in shape and attaches to the cylinder cooling intakes on each side of the propeller hub. The particular model described can be set for low heat and has a thermostat that will control heat during the night. The preheater can be set high for quick preheating, as well. This wonderful device can be purchased through several catalog outlets such as Flyer and Sporty's Pilot Shop. If you order, be sure to measure the amount the spinner on your plane protrudes from the cowling.

Those are just about all the viable choices available for preventing engines from cooling off overnight. They are all effective to various degrees. You can draw your own conclusions as to which would suit your needs best.

Now, let's focus on the various type of quick preheating.

When the temperature drops below 20° F. it is time to give thought to using preheat. Preheating has several advantages beside enabling the engine to start quickly. Foremost of these is warming the oil for lubrication. The $5 to $50 price of an FBO's preheater might sound expensive. Put it in terms, though, of extending engine life and it is a drop in the bucket of a $5,000 overhaul. Warming the cylinders also allows better fuel vaporization which enables the engine to start easily, of course, but it also does two other things. The fuel will burn properly when ignited and, thus, deposits left over from improper ignition will not occur. The other benefit applies to the spark plugs. With a better environment for them to do their work, they will not foul as easily.

Many of the Snow Belt FBO's are ready to wage war on the cold weather. Some of the preheaters are almost as large as a Volkswagen. These are military surplus and are good. Some of them can be rigged up to preheat both engines on a twin at once, in order that the warm engine doesn't cool off while heating the cold one. Now that's an ace of a setup! There are smaller units on the market, though, that serious cold weather flyers can afford.

One preheating arrangement which I would describe as the "bottom of the line" is the hair dryer. The hair dryers that are rated for 1200 or 1400 watts do put out a great deal of heat. The problem is that Mother Nature puts out a great deal more cold at the same time. The heat can be easily directed at the cylinders; but it is the oil pan, however, that needs the most heat. Therein sits a frozen mass resembling a brick—and bricks, my friend, lubricate cylinders and crankshafts poorly.

Hair dryers will work, though. This past winter some pipes froze at my home. They were in a position where using a conventional torch would have set the house on fire. Granted, the pipes would be thawed forthwith, but it seemed safer to use other methods. After a few minutes I decided to use the family hair dryer and fifteen minutes later we had running water again. The ambient air temperature was 8° F. In this case, it was effective. If the temperature is not well below zero, as it sometimes is not, a simple hair dryer will suffice.

Moving up the line of available preheaters is this next arrangement. It uses the heat from your car heater. Sound neat? Let me explain further. There is a unit that fits in a car window and is made snug by rolling the window up against it. This unit consists of a blower or fan that will blow hot air through two ports. On the outside of these ports two hoses similar to vacuum cleaner hoses are connected. The hoses then run between the window of the car and cowling of the aircraft. "How does this connect to the car heater." you ask. Well, it doesn't exactly. The heat in the cabin of the automobile is turned full up and *that* heat is transferred to the aircraft engine.

If this sounds like an inefficient use of energy, you're right. For one thing, no one can sit very long in a car with the heat on high, unless they are naked. Let's face it—when the airport manager pulls up next to you, it's going to be hard for him to understand your idea of an airport sauna.

The cost of this little gem is $230. I kid you not. That's a great deal of bread to lay out, especially when $300 will buy a good propane preheater. But, there is one good point in favor of this system. The blower fan plugs into the cigarette lighter, eliminating the need for an external power source.

One notch above that system is another electric hot air blower. The heat for this model is contained in a hair dryer-like unit. The advantages to this one are that the blower and heating elements are more powerful and the system comes with a stuff hose, that can be shoved into the cowling on one side, and insulative material, that can block escaping heat on the other side. At $190, this system is no great bargain. With a little bit of scrounging or a visit to the nearest Goodwill outlet one could come up with the necessary material to build one as effective. The beauty of it would be the savings in money.

At the top of the line for privately owned preheating systems is the propane-fueled heater. This unit is ignited by a 12- volt battery or external AC power. The attractive point of this unit is the 49,000 BTU heating capacity. This can heat the engine and have the cockpit toasty warm in just a few minutes. To the aviator who flies frequently during winter and wants to avoid the high charges for preheating, this unit is the best type. Ordinarily, these units can be purchased for $265 to $310. At the going rate for preheat service, one could easily make up the cost of the heater in one winter.

Of the methods for preheating an aircraft engine, automobile exhaust is not recommended. This is more than likely the dirtiest air source one could find. Even with today's high EPA standards, the exhaust is full of carbon and carbon monoxide. Despite the possible introduction of carbon monoxide into the cockpit area, there are some advantages—low cost and ease with which it can be rigged up. Connect a heater hose from the tail pipe to the cowling, inlets and you're in business. Once again, I'd like to stress that the carbon monoxide content in the air may or may not generate problems. Beware.

A word about cockpit preheating is in order. Heating the instruments in the panel is the important item. Sure, it's nice to hunker down into a nice, soft, warm seat and blast off; but that really isn't what cockpit heating is all about. All those high-dollar instruments—like the artificial horizon, heading indicator and turn and bank—have greased shafts, bushings and ball bearings. That grease can become nearly brittle in severe cold and that will add wear to the instruments. Without saying, the expected life of those instruments will be reduced by some amount.

Starting

When the temperature sinks to all-time record lows, as it has in the past few years, starting becomes an ordeal. This is especially true when the engine has not been preheated. The procedure for starting is outlined in the aircraft owner's manual and should be followed religiously. It has been devised by the manufacturer through research and experience.

Priming is of utmost importance. Although the procedures many seem excessive many times, adhering to the prescribed number of pumps will result in the fewest problems in cranking the old powerplant to life.

Why are aircraft engines so much harder to start in cold weather than automobiles? The reason is the chemical difference between avgas and gasoline. Gasoline is formulated for better vaporization and, therefore, easier cold weather starts. The formula for aviation petrol cannot change from summer to winter. That would cause problems. So, since airplanes are such mobile creatures and fuel bought in Michigan and then flown to Arizona in January might bubble in the tanks and cause vapor lock, avgas remains on a middle-of-the-road formula. Hence, cold weather takes its toll and we must devise methods such as priming to get those reluctant recips turning.

How Priming Works

On carbureted aircraft engines priming is accomplished by a hand pump on the front panel in the cockpit. This pump is further attached to a priming manifold. The manifold is manufactured of very small tubing of copper or brass-like metal. The tubing is only 1/16- to 1/8-inch in diameter. At the end of the tubing are jets which vaporize the fuel. These jets screw right into the cylinder near the spark plug and below the intake valve. Pumping on the hand pump disperses a fuel mist or vapor into the cylinder for controlled burning (Fig. 3-12).

Some of the mistakes that neophyte pilots make with priming are due to ignorance. Many times—in fact, most of the time—pilots learn to fly in the air weather of spring and summer, in which case

Fig. 3-12. In this photo the priming manifold is in the very center. Fuel is injected just above the spark plug.

they never have an opportunity to use the primer. The primary mistake is to pull the primer out and push it right back in. The priming pump is actually a syringe of sorts. The barrel should be pulled out slowly and allowed to fill. It will make a sucking sound. When this sound stops, the barrel is full and can be pushed back in again. It won't be easy. A full ten seconds might be needed to disperse all the fuel. If this technique is not used, sufficient fuel vapor will not be injected into the cylinder. That may result in frosting of the spark plugs—a problem far greater than the engine not starting (Fig. 3-13).

A mechanic told me, while I was researching the material for this book, of another problem he had seen. Many pilots will come to the point on the checklist where it says "Primer as necessary," and delay. Even though this item may be near or at the end of the checklist, any delay in hitting the starter may allow the fuel vapor to settle in the cylinder and become virtually unavailable for ignition. Although proper primer is applied, the engine might not start. Then you are back to square one—or worse, if the spark plugs are frosted.

Another problem that has been the cause of a few accidents is not locking the primer after use. This results in the mixture becoming too rich and causing the engine to run rough. It might cause the engine to quit, but it is more likely to cause panic in the pilot which could result in an unnecessary precautionary landing.

Fuel-injected engines present fewer problems in cold weather. There are two types of engines in use in most planes today: Lycoming and Continental. The procedures for starting them, when they are fuel-injected, are different. Lycomings are primed by advancing the mixture control and cracking the throttle, which vaporizes the

Fig. 3-13. Be sure to pull the primer handle all the way out and let it fill.

Fig. 3-14. Fuel injectors automatically prime the cylinders. Shown is the injection manifold of a Cessna 210. It is the small tubing.

Fig. 3-15. Keeping the runways clear of snow so that we may fly is a full-time job. This photo was shot in Gaylord, Michigan, in March.

fuel. The mixture control is then closed or put in the idle cut-off position. Continental engines, on the other hand, are primed and started with the mixtures full rich. That makes the whole process simple and it adequately primes the cylinders, too (Fig. 3-14).

To reiterate, fuel injectors automatically prime the cylinders by dispersing fuel vapor. That is their normal function, summer and winter. Follow the instructions in the owner's manual for the particular airplane you fly and the fuel-injected engine will perform admirably on chilly days.

On days when or in locations where preheat is not available, a good aid is automotive starting fluid. It comes in a spray can and is largely an ether spray. The mixture is extremely flammable and volatile. The fluid can be bought at most auto parts outlets and you can almost bank on it's starting your engine on the coldest mornings.

As you can see, there are quite a few things to do on a quick preflight. There are also a great many opinions on what must be done and what can slide. These items depend on the severity of the weather, as much as anything. Use good judgment and always think in terms of *safety* instead of how quickly you can get out of the cold. The items that have been discussed are the points that experienced cold weather flyers deem most important. There are other facets of cold weather operation, yet no other area is more important than getting airborne safely—not even weather (Fig. 3-15).

Chapter 4

Winter's Operational Considerations

Getting the airplane out of the hangar, preheated, preflighted and thoroughly prepared are only parts of the battle we face each gloomy winter day that we fly. There are many techniques involved in actually operating the airplane in cold conditions. In this chapter we will examine those techniques and considerations that lead to a safe performance in a wintry atmosphere.

What Next, After Starting?

It was a nearly exasperating struggle just getting the plane out of the hangar. The snow was packed and icy and it is a wonder that a pratfall didn't break your arm. The only reason your gluteus maximus doesn't hurt is because the cold numbed it long ago. There will probably be a bruise back there tomorrow and your wife will accuse you of being kicked out of the meeting, no doubt, when she sees it.

"About time to jump in the plane and fire her up," you think to yourself. That will be a mistake unless you have already planned ahead.

Whenever snow or ice covers the ground around the hangars or tiedown space, the first precautions with flying the airplane should be taken before starting. I'm not talking about preflight duties—at least, not exactly.

The old adage "keep flying the airplane until it's tied down," begins here. With the airplane positioned with all three wheels on a slick surface of some sort, the brakes may not hold the airplane. Recommended idle rpm's on a cold engine are usually in the neighborhood of 1,000. On a slick surface, 1000 rpm may easily exceed the coefficient of friction of the tires and the plane will move forward.

Positioning the airplane for starting is a key item in safe winter operation. Many times, the surface in front of your hangar or tiedown space will not be completely covered with ice or snow; in this case, simply position the wheels on the patches of available bare pavement. Starting then becomes a normal procedure and the warm up does not include any hidden hazards (Fig. 4-1).

On those days when it is snowing or has recently snowed, there may be no good places to position the airplane to prevent sliding. Precautions must be taken regardless of the conditions. The best thing to do when no bare pavement is near is to face the aircraft in the intended direction for taxi. For instance many aircraft owners keep their aircraft in T-hangars which are usually arranged in parallel rows. The longitudinal axis of the airplane should be turned parallel to the hangars, preferably in the direction of the main taxiway. In the event the plane breaks loose on the ice it should move straight ahead parallel to the hangars and not directly into an adjacent one.

Why not use chocks? Well, there are a couple of good reasons. Most chocks that I've seen around lightplanes are wooden. On hard snowpack or ice, wood will slide. Even under the pressure of the tire it may slide. In fact, on icy surfaces the front chock that is not tied to the rear chock may be squirted out from under the tire.

Another good reason not to use chocks when starting on icy surfaces is that it requires a second person. Often on those sunrise departures, there isn't anyone handy. Starting the engine and pulling the chocks has proven to be a costly mistake. Several stories come to mind. The most popular, though, is about the fellow who owned a J-3 Piper Cub. He had the tail wheel tied while starting. The brakes didn't hold and the airplane was off without him, careening around the airport with this fellow chasing it as in a Buster Keaton movie. Other taxiing planes were forced to give it a wide berth. Finally, it bounced. You guessed it—the plane was airborne without a pilot at the controls. Legend has it that it flew for well over an hour until it died from fuel exhaustion. Did it total the airplane? Well, badly yes, but not beyond repair.

The idea of a plane flying around by itself is amusing. The owner in such a situation probably would not think so, however. The ramifications are numerous. The damage to others' property could

Fig. 4-1. Be sure the surrounding area is clear, especially ahead of the plane, if you are going to start on snowpack like this.

run into millions if it hit the right spot. If the airplane accidentally killed someone, a pilot could possibly face charges of negligent homicide.

The chances of a plane becoming airborne without a pilot are slight. The chances are much greater for it to taxi into a parked airplane nearby or the side of a hangar.

The idea of using a passenger to pull the chocks may be sound in some cases. However, if the passenger is not familiar with airplanes, it may be too dangerous for that individual. It seems that non-aviation people seem to forget the swinging prop because the moving prop is nearly invisible. On this count, I get squeamish; it is just too irresponsible.

Taxiing Technique

One afternoon I drove to the local airport in Gaylord, Michigan. Galord sits on the 45th parallel and is deep in snow country where it begins snowing in October and continues through April. The people who live in that area are used to living with the complications that snow brings. Yet, pilots there are affected by the same diseases southerners are—they contract a "get-home-itis" or "brain-in-the butt" syndrome. As I walked from the parking lot to the terminal, I couldn't help but notice the stiff wind which, at 35 knots, was cutting through everything. I half expected the hangars to be sliced in two.

Inside, the local yokels were having a field day watching a guy in a Cessna 182 trying to get to the runway. It was obvious. Given the conditions of the ramp and taxiways—snowpacked and icy—this fellow who insisted on trying to takeoff had been stricken with "brain-in-the-butt" syndrome. The plane would absolutely not respond to the pilot's inputs. Consistently, the best the pilot could do was to turn the airplane 30° to the wind and, invariably, it would turn back into the wind. The 182 was behaving like the weather cock on a farmer's barn.

Airplanes are built like weather vanes. The rudder cannot help but steer the plane into the wind. When the surface conditions are slick, the tires cannot provide enough friction for steering. In high winds, the plane is at the mercy of nature and one should not be taxiing.

This past winter I had this same problem. It was then that I found the advantage of a turboprop. It was possible to taxi by countering the wind, with reverse on the engine on the side to which we were turning. Turboprops are the only aircraft that have this ability to a high degree.

Even having a plane of great weight provides no advantage when winds are high and the surface is covered with nature's worst. In Alaska several years back, the wind took a Boeing 747 off the taxiway. The behemoth was helpless. Nature was superior to the plane and the captain's skill.

The point is, there are times when no airplane should leave the hangar or tiedown area.

Many times though, old man winter is not so angry. On these days, venturing from the nest can be done safely if we exercise caution. Common sense tells us to go slowly. On icy roads we drive our cars more slowly and the same technique applies to airplanes. When approaching a turn we should slow down well before beginning the turn. Single-engine airplane pilots should not slow down to a snail's pace. The nose wheel seems more effective at speeds slightly higher than a crawl. Taildraggers may need to be slowed down to avoid ground looping. Twin engine airplanes can turn corners more easily by using a squirt of differential power on the outside engine. If the plane turns too much, the other engine can be used to counter the turn.

One mistake in a turn on ice is to use the inside brake to tighten the turn. Characteristically, this will destroy whatever friction does exist between the tire and ice. That, in turn, puts more centifugal load on the outside wheel than the friction there can handle. The airplane then begins to slide to the outside of the turn and there is precious little that can be done to counteract the slide, once it begins, other than releasing the brake pressure. Once the plane slows, normal friction will take over and the plane will begin to turn again. Hopefully, it won't be too late. The best advice is to plan turns well ahead of time and keep your heels on the floor and toes off the brakes.

The art of stopping an airplane on ice is easily learned. As with turning, the key is to plan ahead. It is helpful here to use techniques from driving your car. Gently pumping the brake pedals accomplishes the task properly. Locking the brakes invites uncontrollability. As we discussed above, when the friction is interrupted on the tires, the wind can take over.

One other thing should be mentioned before we leave taxiing technique. The transition from landing rollout to taxi can be dangerous. Don't try to make a turnoff if the plane is moving faster than a crawl. Besides the obvious problems of taxiing that we just talked about, an added factor is present. If you are arriving at a strange airport or have not been there in the last 30 minutes, conditions may be other than what we expect. At strange airports, crosswinds may be present that are misdirected from hangars, buildings or surrounding terrain. So take it real slowly until you have a grasp of the situation.

If you were just on a local flight over the area ski resorts, conditions may have changed somewhat since departure. Again, go stowly until the situation is familiar. Take your time taxiing back to the hangar. After the flight is over is no time to wreck your airplane.

Takeoff Technique

The techniques for takeoffs on icy surfaces are pretty much what they are in the summer. If you are adept at takeoffs during warmer times, winter should give you little problem. Simply put, this phase of flight stays most nearly the same all year around.

There are some considerations that need to be examined. If the runway is icy or snow covered, multi-engine aircraft are affected the most. A landing strip that is quite adequate in warm weather can be much too short in the cold. I am referring to the accelerate-stop distance—that is, the distance the plane covers accelerating to Vmc (minimum single-engine controllable velocity) and stopping as if engine failure occurred at Vmc. If the runway is snowpacked, the distance to stop will be increased by the slick surface. A good safety

Fig. 4-2. This takeoff will be a snap for a change. They got the runway cleared!

factor would be to increase your accelerate-stop distance by 10 to 20 percent. This should be a good margin (Fig. 4-2).

If conditions happen to be slushy or covered by four and five inches of new fluffy snow, the accelerate-stop distance is affected again. This time it will be the inability of the airplane to accelerate to Vmc. This type of condition affects the "singles" as well. If the strip is short to begin with, an airplane may not be able to attain flying speed before reaching the departure end of the runway.

In slushy conditions, stopping becomes difficult. One reason is that the slush tends to grab onto the wheels, so to speak, and exert a force that may change the plane's direction of travel. This may set up a hydroplaning situation, which may also occur during acceleration to takeoff speed.

There are techniques for counteracting the effects of water, slush and snow on the runway. Basically, we can expect an increase in takeoff distance or we can limit our gross weight. In the next section we will examine some FAA tests on the subject.

In slush or fresh snow, the best method is to use a soft field takeoff. Keep the elevator all the way back to the stop to pull the nose wheel up and out of the slush. Two things are accomplished by doing this. Drag is reduced on tricycle gear airplanes because only two wheels are in the muck and a high angle of attack is achieved. Disregard the stall warning horn on soft field takeoffs. The airplane will not lift off the ground if it isn't ready. True, the airplane is in ground effect and will fly at a lower speed. If the airplane is flown out of ground effect at the same angle of attack at which it lifts off, it *will* stall. The remedy is to level off in ground effect as soon as the main wheels break loose of the muck. (Ground effect is roughly equal to one wingspan of your aircraft above the ground.)

Water, Slush and Snow on the Runway

In 1960 and 1961, the FAA ran some tests that uncovered some interesting facts. Though they apply to heavy transport turbojet aircraft, parallels can be drawn. That is why they are included here.

Early in the operation of turbojet aircraft it became evident that correction factors should be applied to takeoff data in order to keep the aircraft performance within safe parameters when water, snow or slush was on the runway. The first test used a Boeing 707 with a slush depth on the runway of 6/10-inch. The test showed that retardation of acceleration on takeoff was of such consequence that gross weight had to be limited for a critical field length.

In August 1961, further tests were conducted at the National Aviation Facilities Experimental Center (NAFEC) by the FAA/NASA. (Don't ask me where they found slush in August.) The test used a Convair 880 which is very much like a Boeing 707 in appearance. The object of these tests was to obtain data regarding the retardation effects of slush and the effects of hydroplaning on takeoff performance. In addition, they investigated control problems and damage induced by the slushly environment.

The tests at NAFEC were conducted on a slush-covered 10,000-foot runway. The slush ranged in depth from zero to two inches and aircraft velocities were 80 to 160 knots. The retardation forces that were measured from deceleration data were considera-

Table 4-1. Runway increases developed by airlines.

TYPE AIRCRAFT	TAKEOFF WEIGHT	REQUIRED RUNWAY INCREASE (APPROXIMATELY)
Douglas DC-8	251,000	10%
Boeing 707/100 Series	247,000	15%
Douglas DC-8	296,000	14%
Boeing 707/300 Series	296,000	15%
Convair 880/22M	150,000	15%

bly higher than those predicted from wheel and tire drag tests run earlier. Theoretical studies which neglected the factors of slush spray impingement and hydroplaning were found to be excessively in error. The impingement of slush against the aircraft and landing wheels contributed tremendously to drag forces. At velocities above 120 knots, the aircraft began hydroplaning. One good effect, however, was that the hydroplaning lowered the drag forces. Based on the information from these tests, the FAA formulated these guidelines:

a) Takeoffs should not be attempted when water, wet snow or slush is greater than 1/2-inch on an appreciable part of the runway.
b) In operating in depths of the wet stuff up to 1/2-inch, correction factors should be applied.

As is so characteristic of the FAA, they did not release any data on just what correction to add. Operators of the airlines took it upon themselves to take the data that were available and derive some fairly good operating limits.

The chart is used in this manner: *The required runway increase is multiplied by the dry runway requirement at a particular gross weight.* For instance, a DC-8 at 251,000 lbs. needs a runway of 6,500 feet for a dry day takeoff. Using the 10 percent increase, we add 650 feet to the 6,500-foot figure and find a total of 7,150 feet needed for takeoff on a 1/2-inch slush covered runway (Table 4-1).

Another way to tackle the same problem is to reduce gross weight. Some commercial operators have arrived at the correction factors shown in Table 4-2.

The weights shown are takeoff weights for dry runways. The weight reduction factor is then applied whenever slush covers the runway. For example, a Boeing 727 at 152,000 lbs. gross weight on a dry runway must limit its gross weight to 132,200 lbs. for a 1/2-inch of slush.

The parallel we can draw from these figures is that all planes are affected by 10 to 15 percent. As a result, it would be good to use the

Table 4-2 Gross weight reductions developed by airlines.

TYPE AIRCRAFT	TAKEOFF WEIGHT	WEIGHT REDUCTION
Boeing 727	140,000	–17,500
Boeing 727	152,000	–19,800
Boeing 720	180,000	–10,000
Boeing 720	190,000	–11,000
Caravelle	110,000	–11,000

airlines as a model and also use a factor of 10 to 20 percent for smaller planes. Airplanes with low horsepower will have the greatest problem in getting airborne. Also, aircraft such as Beech KingAirs and 99's have five or six tires on the ground which add extra drag. Thus, the 15 percent factor for such airplanes is most realistic.

Hydroplaning

Hydroplaning can occur on either takeoff or landing whenever sufficient speed exists. Hydroplaning can be considered potentially dangerous all the time because there is lack of both control and braking efficiency.

Hydroplaning is caused by a film of water on the runway. As the speed of the aircraft and the depth of the water increase, the water layer builds up an increasing resistance to displacement. This results in the formation of a wedge of water beneath the tire. The vertical component of the resistance progressively lifts the tire. Less tire area is then in contact with the runway and controllability decreases. At some point, the tires will fall completely out of contact with the runway, no longer contributing to aircraft control and rendering braking action nil. Except for the rudder, which may not be very effective, the pilot has no control over the direction the airplane decides to travel. In fact, the pilot becomes a passenger.

There are three types of hydroplaning:

Dynamic hydroplaning occurs, as described above, when about 1/10-inch of water covers the runway.

Viscous hydroplaning involves a thin film of fluid, not more than 1/1000-inch in depth which cannot be penetrated by the tire. Instead, the tire rolls on top of the film. This can occur at much lower speeds than dynamic hydroplaning, but it requires a smooth or smooth-acting surface.

Rubber Hydroplaning requires a prolonged locked wheel skid, reverted rubber and a wet runway surface. The reverted rubber acts as a seal between the tire and the runway. Water, delayed exit from the tire footprint, is then heated and converted to steam. Hence, the steam supports the tire off of the runway.

It is possible for us to predict, with accuracy, the speed at which the aircraft we fly will begin hydroplaning. Data derived from FAA tests revealed the minimum dynanic hydroplaning speed of a tire to be 8.6 times the square root of the tire pressure. I use

$$9\sqrt{\text{Tire Pressure}}$$

because it is easier to remember and close to the same figure.

For example, a Jet Commander 1121 has a main tire pressure of 45 pounds. The calculated hydroplaning speed is approximately 100 knots. For a Cessna 172, the dynamic hydroplaning will start at about 50 knots.

Many times, the nose tire pressure is lower than the mains. This means hydroplaning will begin at a lower speed on the nose tire. For high speed turnoffs on wet runways, the airplane should be slowed below that speed before starting the turn.

One problem with hydroplaning is that once it begins it may continue at slower speeds than the calculated speed of its beginning. Thus, it is very important to slow down well below the calculated values.

Landings

A friend of mine had been out in a Cardinal RG. When he returned to the home base, the wind had become stronger by several knots. Mike is a great pilot, but sometimes circumstances arise that turn a pilot into a passenger. The runway was completely snowpacked and could have matched the scenery in any number of northern areas from New York to Alaska. Ordinarily, snowpacked runways won't give a person that much trouble unless there is a stiff crosswind. On this particular afternoon, the crosswind was bountiful.

A tough thing about flying is recognizing when a crosswind is too much for the plane one is flying. Almost every airplane has a demonstrated crosswind component (Fig. 4-3). In fact, the one I fly professionally has two: one for 50 feet of altitude and one for six feet of altitude. The demonstrated crosswind component is the amount of direct crosswind test pilots have found to be limiting with maximum control inputs. (Piper products post, or placard, these for each of their planes, which is a tremendous help. On this count, Cessna is light-years behind.)

As Mike turned final for the east-west runway, the approach was normal. Over the threshold, the Cardinal whistled. He kicked out the crab and tilted the upwind wing for the side slip. The upwind wheel touched first—perfectly executed crosswind landing. The plane began slowing and the rudder pooped out. The runway was extra slick due to a little melting of the snowpack. Cocking into the wind, the airplane did just what it was supposed to do and not what Mike wanted it to do. Now sliding sideways and cross-downwind, the plane smoothly made its stop. The tail was only inches from the plowed snowbank.

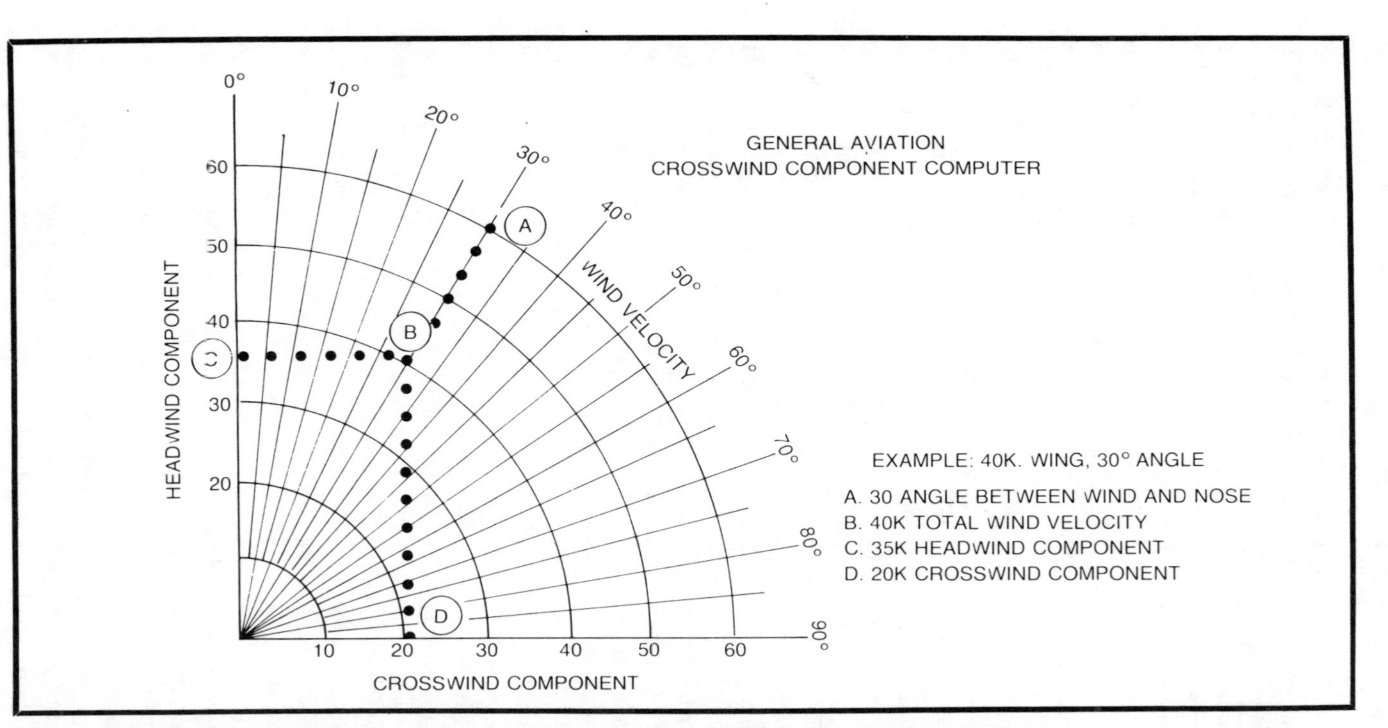

Fig. 4-3. A handy piece of equipment is the crosswind computer. Use it well before you get to final approach.

The plane was undamaged, but it had to be towed into the hangar. There was too much wind and not enough friction to taxi. In conditions such as those, it is impossible to predict whether that may happen on rollout or not. One thing is for sure: if the demonstrated crosswind component for your aircraft is 26 knots, for example, the plane may slide if the rudder loses effectiveness under 35 knots. Most general aviation aircraft will lose the use of the rudder between 30 and 35 knots. On a slick surface at or near the demonstrated crosswind component of your plane, figure on sliding.

Once a slide or skid has begun there isn't much to do but ride it out. To minimize the effect or results of a slide, the landing should be made as near the upwind side of the runway as possible. The danger here is catching a wingtip in a snowbank. The benefit is if a slide occurs, we have maximum distance to allow the airplane to come to rest by itself.

In some cases, a local snow shower will move through an area and leave the runway covered with new fluff. If a landing is made in near one-mile visibility, the bottom becomes hard to find. It is something akin to the floatplane landing on glassy water. The best way to handle this situation is exactly the same way a floatplane pilot would. Set up a 100 to 150-feet per minute descent rate well out on final approach. Continue the rate till touch down. The landing may not be silky-smooth but the alternative is dropping it in from ten feet because one's judgment erred in the lousy conditions.

Also, in conditions of new fluffy snow, soft field landings become the order of the day. There are a couple of schools of thought about how to make the approach as far as the use of flaps is concerned. One method is to use full flaps, hang it on the prop and touch down at the lowest possible speed. The other method is to use partial flaps, such as 20 to 25° and opt for a higher touch down speed and, thus, a low rate of descent. This is the method I endorse because planes flown with full flaps will stop flying suddenly at the stall. Planes with partial flaps will stall also, but the attitude for a soft field landing is nose-high. It is easier to transition from the 20° flap setting to a nose-high attitude. With full flaps, the transition to the traditional soft field attitude is more difficult, partly because of the extreme nose-down pitch associated with full flaps and partly because of the extreme angle of attack for that configuration needed for touch down.

In other words, flying the airplane with a nose-high attitude is the best in the author's opinion. Whenever landing in mud, slush, snow or a combination of these, the main wheels tend to stick (drag)

Fig. 4-4. Flying off cf skis has been around for a long time. Check this Great Lakes on skis about 1930.

on touch down. That forces the nose down from the landing attitude, at least in tricycle airplane. If a slightly higher touch down speed is used, elevator control is prolonged; thereby, the nose can be held up out of the slop for a longer time.

In taildraggers, using partial flaps is still the best choice. The choice between wheel landing or three point is up to the pilot. A compromise may be in order, though. If the nose is kept high, the prop will be less likely to be nosed into the slush or snow. If the nose is kept high for a three point, the plane may drop in at an inopportune moment. The higher speeds seem to work the best, as all control surfaces remain effective until the plane is tracking down the runway.

Short field landings in the winter don't present much of a problem if the surface is slushy or snow-covered. The slop will slow the plane down quickly. If the runway is snowpacked, then touch down with full flaps near the upwind side if there is a crosswind.

After landing, gently slow the aircraft down. Don't try to make the first turnoff you approach. The airplane should be down to a slow walk in icy conditions. Always remember that friction is not there as it is in summer. The plane may handle well straight ahead and lead a pilot to forget the surface he is taxiing on. Many accidents happen every winter when pilots let their guards down.

Flying Off Skis

Flying off skis makes the airplane the ultimate escape vehicle. It is for the outdoorsman; remote fishing and camping spots are at hand in the winter. One can enjoy the wilderness experience right through the winter. But flying on skis is a specialty and requires a check-out from a bonafide ski-plane instructor (Fig. 4-4).

Flying an airplane equipped with skis is different. The airplane reacts much more like a floatplane than a wheel plane. Wind is a critical factor because a ski-plane, like a floatplane, will weathercock into the prevailing wind. A tricycle gear ski-plane, because of its elevated tail, tends to weathervane more than a taildragger. Also, flying with skis means flying or maneuvering without brakes. That means wind and velocity are primary concerns to safety before the airplane even moves.

Preflight checks on a ski-plane involve the usual items. The skis need close inspection. The clamping bolts that attach the skis to the airplane, the cables and shock cords are items on the list. The retracting mechanism on retractable skis should be checked for fluid and leaks. The limiting cable is the most important item. If it fails, the tip of the ski will hang down and make landing interesting, to say the least.

The tail wheel must be checked thoroughly for cracks or signs of failure. Without a tail ski, the tail wheel assembly takes a great deal of punishment. The usual technique is to taxi at a high enough speed to elevate the tail. The tail is lowered to add drag for the stop.

For flying deep into north country, anti-ice fuel additives are recommended. However, you should check with the General Aviation District Office (GADO) as to which additive to use.

Some skis are clamped on or roll-on or a combination with the wheels. On these types, the tire pressure should be checked. Tires can be expected to expel 5 percent of their air on a daily basis. Also, moving the plane from a warm hangar to the frigid outdoors will cause a pressure drop in the tires.

Some other things to be considered for an overnight trip are de-icing fluid and survival gear. Check that these items have been stowed. Wing covers are also great to have whenever there isn't hangar space available.

Engine run-up on ice or snow takes special consideration. On heavily crusted snow or glare ice, the engine run-up is best done in the parking area. Be sure the area behind is clear. The prop will throw chunks of snow to the rear. The engine should be good and warm before doing the run-up to promote longer engine life. The parking area makes an ideal run-up area, though, because the skis are usually frozen to the ground and often it takes full power to break away.

On glare ice, the plane will need to be left tied down. Other alternatives are to place a rock or sand bag in front of the skis. The latter works the best.

With the exception of crusted snow and glare ice, ski bottoms freeze easily and become difficult to free. The friction heat that is generated during taxi will melt the snow underneath when the airplane comes to rest. The result is a strong bond between earth and plane. Because of this problem, preplanning is essential. This is especially true when in a remote area where extra help is scarce. The best thing to do is pull the skis up on tree boughs or burlap bags.

Whenever the plane is stuck to the snow, several helpers and full power will shake it loose. Flaps sometimes help. If a pilot is alone, then full power and working the elevators and rudder may wrestle the airplane loose.

Hydraulic retractable skis can be pumped up and down when stuck. This is not recommended, however, because hydraulic lines and seals can rupture as a result of the extra pressure on them.

Whenever the skis do break loose, the trick is to keep moving. The friction generated from taxiing will heat the skis slightly once more and refreezing will take place quickly once you stop.

When learning to maneuver on skis it seems quite clumsy and plenty of room is necessary. (A couple of counties ought to do it!) All the directional control comes from either the nosewheel or a steerable tail wheel and the propwash across the rudder. Floatplanes have difficulty in making downwind turns due to the tendency of the plane to weathervane into the wind. The ski-plane is affected the same way. Forward elevator pressure and high power give the necessary control. The idea is to lighten the tail without nosing over into the snow.

The sailing technique that floatplanes use is most helpful in taxiing in a brisk tail wind. In either a taildragger or tri-gear plane the elevator control should be turned away from the wind. Simply, if the wind is a quartering right tail wind, the controls should be forward and full left. This puts the aileron down on the right wing and the wind has a sail to blow on. The tendency to weathervane is thereby decreased.

Another good tip is to open the door on one side or the other of the plane to act as a sail. This only works well on tri-gear airplanes, however.

Skidding while turning is natural in a ski-plane. Planning the projected track is important. Plan for enough room and how to miss drifts, ridges and hummocks. In strong winds on icy surfaces, it is best to use wing walkers on the tips while taxiing and lining up for departure.

Once again, ski-planes tend to weathervane into the wind. The sailing technique is the best thing to minimize drift tendencies. *On ice or crusty snow,* power and speed should be kept to an absolute minimum. On the other hand, *in new powder,* "step taxiing" (borrowed from floatplanes) means taxiing just below takeoff speed. The higher speed is necessary to keep the plane from sinking into the soft snow and becoming mired.

Glare oftens affects depth perception. In many areas, airport managers have the snowbanks painted with a red stripe to aid in perception. Sometimes tree boughs are pushed into the snowbanks and this works well. Without such aids, a taxi accident is more probable.

Snow adversely affects takeoff distance. In wet snow the distance required may be two or three times the normal distance. On ice, such as crusty snow or a frozen lake's surface, the distances need only be lengthened by 100 feet over the flight manual's figures. For normal snow conditions, where depths are not excessive, 1000 feet is a comfortable amount to add to takeoff data.

Experts recommend partial flaps during taxi and a soft field takeoff technique should be utilized. Remember to keep the airplane moving and scribe wide arcs to minimize the tendency to ground loop due to weathercocking. When lining up for the takeoff, keep the plane moving as it can quickly settle and become glued to Mother Earth.

If the snow is deep, sufficient airspeed for takeoff may not be achieved. In this case, make several runs down the same tracks to pack the snow. Then depart in your tracks. If this doesn't work, an adjustment in the gross weight may be necessary. As you can see an airplane with a STOL conversion would be an asset.
be the rule of the day on skis. The reason for soft field takeoffs on wheels is to get the wheels out of the slop as quickly as possible. That reason is good for ski-planes as well, with a few hidden subtleties. Glare is always a problem, even on cloudy days in the snow. That glare can hide a multitude of evils such as drifts, rough ice, windrows and soft sticky spots. For these reasons, getting airborne quickly is smart.

Crosswind takeoffs are quite different on skis. Throughout this section we have compared skis to flying floats. Crosswind techniques don't compare between these two types of flying, however, because floats have water rudders and skis have no such counterpart. Therefore, during crosswind takeoff runs, the plane may crab (on the ground) or weathercock. This, in itself, adds a great deal of drag and increases takeoff distance.

Cruise speeds on planes equipped with skis are lower. This is partly due to the increase in weight and partially to drag. The installations that keep wheels on the plane, such as retractable skis or wheel-through-the-skis set ups, give the greatest penalty. Cruise speeds may be 30 knots lower than normal. When the wheels are absent, there isn't too much speed difference—maybe only 3 knots or so. Retractable skis and mechanisms represent weight increases, so remember to include that extra weight in the weight and balance figures.

Landing a ski-plane is the easiest kind of landing. It is so easy, in fact, that some pilots have to be re-trained for wheels when spring arrives. Whenever operating off airports, though, certain precautions must be taken to escape hidden hazards.

Landing a ski-plane is more than spotting your fishing hole and putting it down. Several passes over the area are needed to assess the situation. The items to check for are the length available in the direction of the wind and location of obstructions.

Under bright conditions, depth perception can be extremely impaired. Yet, a soft field landing is necessary for survival of the aircraft, so it is also a good idea to land near a row of trees or shoreline so some contrast is available. Because the depth of snow is unknown, an abrupt stop or deceleration could occur. Expect the worst.

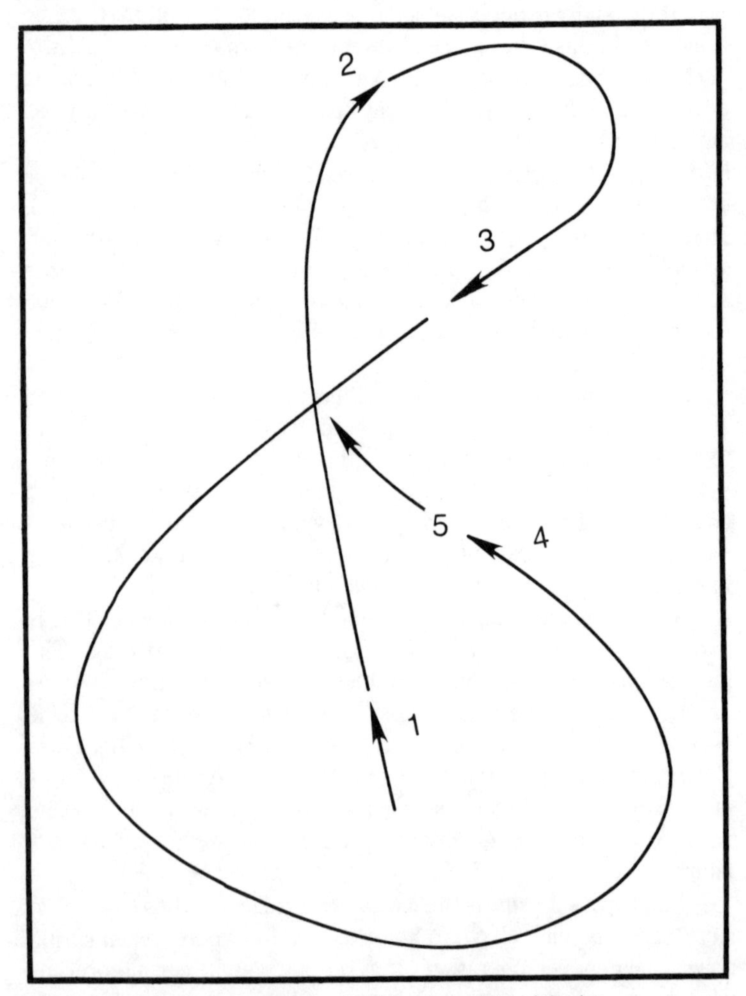

Fig. 4-5. The figure-8 method for checking the safety of the ice.

Snow-covered lakes are a favorite rendezvous point for ice fishermen. Caution is urged around the downwind end of these lakes. Deep drifts often accumulate there and ski-planes have been half buried when pilots unexpectedly taxied into them. The best place to choose to land on a snow-covered lake is on the lee side of an island. In this area the snow will normally be shallow and ice smooth.

Trial landings are a good idea (see Fig. 4-5). By making a trial landing as shown in the figure, hazards are minimized. First, touch down at near takeoff speed. Make a slide for 200 or 300 yards followed by a go around. The next pass will reveal the snow depth.

On the second landing we begin the "figure 8" maneuver. Maintain speed near rotation speed. After sliding a bit turn either right or left and cross back through the original tracks. At that point check for a dark discoloration. If it is dark, then water is on top of the ice. Continue the maneuver and circle in the opposite direction. Intercept the original tracks and stop. Now check the ice thickness and quality.

Blue ice is a good indication of sufficient ice thickness. Dark patches may indicate extremely thick areas. Gray-looking snow usually means the ice has water on top of it.

The best policy is to land on lakes instead of rivers. Lakes don't have the under-ice currents that rough the surface up and the ice is of a more uniform thickness. Rivers tend to have rough ice near the shoreline due to currents. Many things affect ice safety and stability: currents, ambient air temperature, resonance waves and resulting cracks, and water temperature and depth. Always land as near the shore as possible because the ice can be trusted more there.

The following table is a good guideline to ice thickness.

Table 4-3. Recommended Ice Thickness for Ski-Equipped Aircraft.

AIRCRAFT WEIGHT	FRESH WATER ICE AIR TEMP. DEGREES F.			SEA ICE AIR TEMP. DEGREES F.		
	14°	22°	31°	10°	19°	28°
2400 lb	5"	6"	7"	9"	11"	13"
3000 lb	6"	7"	8"	10"	12"	14"
5000 lb	8"	9"	10"	12"	15"	17"
7000 lb	9"	10"	11"	13"	16"	19"
8000 lb	9"	10"	11"	14"	16"	20"
8500 lb	10"	11"	12"	15"	18"	22"
9500 lb	10"	12"	13"	16"	19"	23"
SUBJECT TO VARIABLES SUCH AS AIR TEMPERATURE, STRUCTURE, CRACKS UNDERLYING SUPPORT, RESONANCE, NUMBER OF LANDINGS AND PARKING TIME.						

The fact that flying is possible because of machins means there will be mechanical problems. Skis are susceptible to certain problems. Retractable skis give the most problem, simply because they are the most complex. Ordinarily the glitch is an incomplete extension. The gear should be checked during preflight and most problems can be found then. If they are not found, a partial extension of the ski will show a furrow from the tire in the track. Also characteristic of this eventuality is extra drag. The plane will pull strongly in the direction of the partially extended ski. The problem should be corrected before the next takeoff, if at all possible.

Limiting cables cause problems that should be caught on preflight inspection. When the forward cables become unattached for some reason the ski toe will hang down. The ensuing landing is touchy, to say the least.

Something the novice ski-plane pilot may not realize is that parking on ice for an extended period requires greater ice thickness. For up to six hours under moderate 19 to 22° temperatures, an increase of 25 percent should be added to the required ice thickness table. Another tip: don't park near cracks. That is harder than it sounds, since ice is more often than not covered by snow.

Once again, flying skis is a great deal like flying floats. A thorough check-out is necessary or a mistake can easily be made. Another sound idea is to make several trips with a seasoned ski pilot. One could learn more there than in reading all the books written on the subject.

Carburetor Ice

The carburetor ice to which reciprocating engines are susceptible was a probable cause, or a factor, in 360 general aviation accidents which occurred during a recent five year period. In those accidents there were over 40 fatalities and 160 injuries out of a total of 636 pilots and passengers.

Although carburetor ice can occur in summer, cooler temperatures are more conducive to the formation of induction icing. In the following few pages the term "carburetor ice" will refer to any type of induction ice that affects reciprocating engines.

Carburetor icing accidents can be attributed to the pilot in virtually all cases. As a result, improved pilot awareness and caution should reduce the incidence of carburetor icing accidents.

Awareness of carburetor icing conditions sometimes requires extra effort because of unique combinations of engine installation and operation and weather. Familiarity with certain general information,

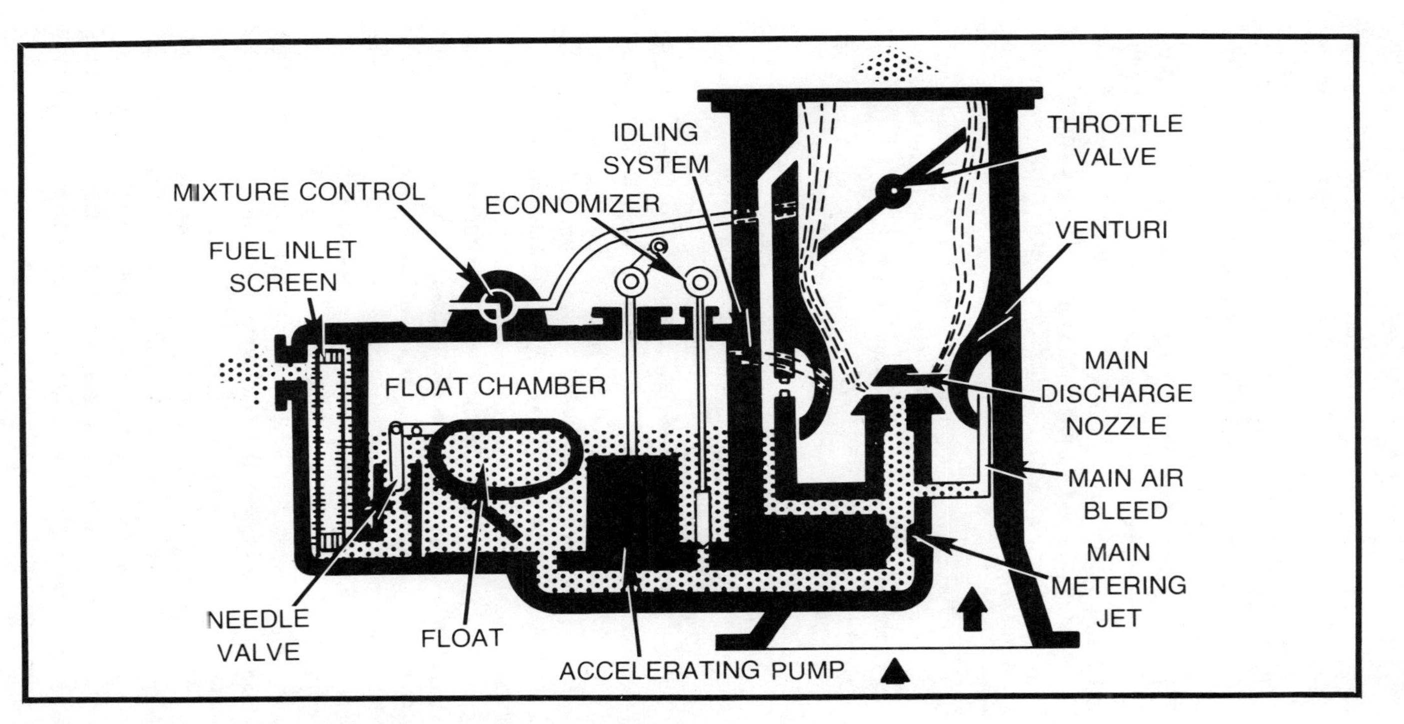

Fig. 4-6. The typical float-type carburetor found on so many lightplanes.

coupled with information on the specific airplane one is flying, will sufficiently arm a pilot with the knowledge necessary to combat carburetor ice.

It is important for pilots to know the three categories of carburetor ice. These categories are impact ice, fuel ice and throttle ice.

Impact ice is an appropriate name. This type of carburetor icing *is* typical of the wintertime environment. It is formed by the impingement of moisture-laden air at temperatures between 15° and 32° F. onto elements of the induction system. Those elements such as the air scoop, heat valve, carburetor screen, throttle and carburetor metering elements need to be below 32° for the ice to stick (Fig. 4-6).

When the ambient air temperature is in the 25° F. range, conditions are prime for impact ice. The moisture in the air is in a super-cooled state and only needs a surface (such as a moving plane) to disturb the equilibrium to precipitate ice development. Hence, pilots must be aware of this condition when operating in snow, sleet, rain or clouds.

The best remedy for impact ice is a change of altitude. Also, a switch to the alternate air source may become necessary if the conditions persist and cannot be avoided.

Fuel ice is a type of carburetor icing that is not necessarily associated with cold weather. Fuel ice forms downstream from the point at which fuel is introduced. If there is sufficient moisture and if the temperature in the carburetor has been sufficiently cooled by fuel vaporization, the moisture will condense. This cooling process takes place in the aircraft induction system when the heat necessary for fuel vaporization is taken from the surrounding air, thus cooling the air. It is a known fact that cooler air can hold less moisture than warmer air; so whenever the air is cooled, some amount of the moisture is precipitated. If vaporization cools the surroundings sufficiently, the condensate will freeze. Any structure that impedes a straight-line flow (such as an adapter elbow) and lies in the path of the moisture as it freezes, will initiate ice accretion at that point. If no anti-icing action is taken, the ice buildup will eventually throttle the engine.

Visible moisture such as snow or rain does not need to be present in the air for the formation of fuel ice. Only air of high humidity is necessary. The fact is that fuel ice can occur on the brightest, sunniest warm days. Therein lies its insidiousness. Many pilots have been taken in because the day didn't look like one when ice would form.

The usual range of temperature at which fuel icing may be expected to occur is from 40 to 80° F. The upper limit of this range can be as high as 100° F., but the optimum temperature is 60° F. Humidities must be 50 percent or above. As the humidity climbs, the possibility for fuel ice formation increases.

Fuel ice is not a problem in planes that use fuel injectors since the fuel is injected into the engine at a point beyond any areas where the temperatures may be freezing. The fuel is dispersed directly into the cylinders where it is generally warm, except at first start. Engines with centrifugal superchargers introduce fuel downstream of the impeller where splash-back on the impeller blades would cause fuel icing.

Throttle ice is formed around a partially closed throttle when water vapor in the induction air condenses and freezes due to the expansion cooling and lower pressure as the air passes the construction made by the throttle butterfly valve. The temperature drop normally does not exceed 5° F. Therefore, when ambient air temperature is above 37°, a pilot need not worry about the formation of throttle ice. At least, this is true if only air is passing the throttle such as in a fuel-injected arrangement.

When there is a fuel-air mixture at the throttle, ice formation may become a reality at temperatures much higher than 37°. The fact is that fuel ice and throttle ice may very well combine to choke the engine. One engineering design that eliminates throttle ice is location of the throttle in a warmed region such as between a turbocharger and the cylinder.

Simply stated, any of the ice forming situations above can cause a loss of power by restricting the air injected into the engine or interfering with the proper air-fuel mixture. It is smart to use carburetor heat as an anti-icing agent. Waiting until carburetor ice is suspected may result in a vicious circle. If the engine is producing less power, then less heat is available to apply to the carburetor to melt ice, and so on. Therefore, it is wise to guard against buildup of carburetor ice before de-icing capability is lost.

In order to get a better grasp of carburetor icing that might be expected in lightplane systems, let's look at a test report from the FAA.

"Two typical lightplane installations were tested, one with a float-type carburetor, the other with a pressure-type carburetor. With the first, serious icing occurred up to carburetor air temperatures of 62, 63, and 93 degrees Fahrenheit, and the lower limits of relative humidity of 80, 60 and 30 percent, for high-cruise, low-cruise, and glide-power conditions, respectively. With the pressure-type carburetor installation, the results were serious icing between carburetor air temperatures of 48 and 55 degrees Fahrenheit with relative humidity from 90 percent to 100 percent at low-cruise power, and up to approximately 75 degrees Fahrenheit with relative humidity greater than 32 percent at glide power. No serious icing occurred at the high-cruise power condition."

Carburetor heat in small aircraft is generally derived from an exhaust pipe cuff. That hot air is directed into the carburetor intake, as needed, by a valve. When full carburetor heat is selected, the outside air intake is closed.

One should realize that partial carburetor heat can be worse than none under some conditions. For example, if the fuel-air mixture had a temperature of 20° F., ice might not form. If partial carb heat was applied and raised the fuel-air mixture to 30°F., ice formation would be more likely. Full heat would raise the temperature completely out of the freezing range. It is unfortunate, but most light aircraft with small engines don't come equipped with a carburetor air temperature gauge. As a result, application of full carburetor heat is the best bet. On aircraft with larger engines, turbocharging or supercharging, more discretion is needed in using carburetor heat because excess heat may cause engine overheating and detonation. As a rule, temperature instrumentation should be installed on airplanes of this kind to assist the pilot in modulating the appropriate amount of heat.

After talking about the importance of using carburetor heat when necessary, we must now discuss overuse. Carburetor heat can be overused, resulting in lower power and higher cylinder temperatures and which, at times, can be critical. A go-around, for example, calls for full power, but with the carburetor heat left "on" a loss of power results. When ambient temperatures are high, the misuse of carburetor heat can cause cylinder overheating and possible detonation damage. As stated in the FAA test results we examined a few paragraphs ago, high power settings rarely require carburetor heat.

Here is a bonus tip for cold weather flyers. As it has often been said, "There are exceptions to all rules." In extremely cold and dry weather, with no icing potential, the use of a little carburetor heat may actually increase power to some degree because of improved fuel vaporization. This condition only rarely happens and then, most likely, only in the far northern localities.

Another tip on carburetor heat is to use it during engine warm up. The increased induction temperature will aid in heating the cylinders more quickly.

Appraisal of carburetor icing potential is tough, to say the least. Induction temperature instrumentation makes the task much easier. So many of us, though, don't have carburetor temperature gauges and, therefore, must rely on knowledge of temperatures of the ambient air and the relative humidities. With relative humidities in the heighborhood of 50 percent, we should expect the possibility of ice. A good idea is to check with "flight watch" as we travel along to check the humidity in the area we are transiting.

There are several ways to identify carburetor icing problems. Icing should immediately be considered the cause of a power loss. One thing for sure, carburetor heat is easier to reach for than looking under the cowling in mid-air. On fixed-pitch propellers, a power loss is indicated by a loss of engine rpm. An engine with a constant speed prop reflects a loss of manifold pressure.

Another subtle hint of carburetor heat would be a slight nose-down attitude caused by the loss of power. Upon trimming the nose back to level flight, a drop in engine rpm and airspeed will be noticed. This is all assuming a fixed-pitch propeller.

Finally, carburetor ice may cause engine roughness. Sometimes, though, the roughness may not show up until the engine is about to quit altogether. By that time, a pilot should have recognized other symptoms.

Susceptibility to carburetor ice varies from plane to plane. An engine employing a float-type carburetor and having the fuel introduced upstream from the throttle valve would be the most susceptible arrangement to carburetor icing problems. At the other end of the spectrum is the engine with fuel-injection. Only the most troublesome fuel icing will affect these types. However, fuel-injected engines are still susceptible to impact and throttle icing.

The FAA has drawn up several prevention procedures. They are listed below in the FAA form. Every cold weather and warm weather flyer should find them helpful.

"Carburetor icing troubles can be avoided by practicing the following procedures:

1. Periodically check carburetor heat systems and controls for proper condition and operation.
2. Start engine with carburetor heat control in the "cold" position, to avoid possible damage to the carburetor heat system.

3. As a preflight item, check carburetor heat availability by noting heat "on" power drop.
4. When the relative humidity is above 50% and the ambient temperature is below 80 degrees Fahrenheit, use carburetor heat immediately before takeoff. In general, carburetor heat should not be used during taxi because of possible foreign matter entry when intake air is unfiltered in the "alternate" or carburetor heat "on" position.
5. Conduct takeoff without carburetor heat unless extreme carburetor icing conditions are present, when carburetor heat may be used if approved by aircraft manufacturer, and when conditions are such that there will still be ample power for takeoff without incurring engine overheat damage.
6. Remain alert after takeoff for indications of carburetor icing, especially when the relative humidity is above 50%, or when visible moisture is present.
7. With supplemental instrumentation, such as a carburetor air temperature gauge, partial carburetor heat should be used as necessary to maintain safe temperatures to forestall icing. Without such instrumentation, use full heat but only intermittently if considered necessary.
8. If carburetor ice is suspected of causing a power loss, immediately apply full heat. Do not disturb throttle initially, since throttle movement may kill engine if heavy icing is present. Watch for further power loss to indicate effect of carburetor heat, then rise in power as ice melts.
9. In case carburetor ice persists after a period of full heat, gradually move throttle to full open position and climb aircraft at maximum rate available in order to obtain greatest amount of carburetor heat. If equipped with mixture control, adjust for leanest practicable mixture, (approach this remedy with caution—although carburetor ice generally serves to enrich mixture, the reverse can be true; if the engine is lost through excessive leaning, an airstart might be impossible with an iced induction system).
10. Avoid clouds as much as possible.
11. In severely iced conditions, and when equipped with mixture control, backfiring the engine can sometimes be effective in dislodging induction system ice. With carburetor heat control "off," lean engine while at full throttle (observe caution note in No. 9 above).

12. Consider that carburetor icing can occur with ambient temperature as high as 100 degrees Fahrenheit and humidity as low as 50%. Remain especially alert to carburetor icing possibilities with a combination of ambient temperature below 70 degrees Fahrenheit and relative humidity above 80%. However, the possibility of carburetor ice decreases in the range below 32 degrees Fahrenheit. This is because of lessened humidity as the temperature decreases, and at around 15 degrees Fahrenheit any entrained moisture becomes ice crystals which pass through the induction system harmlessly. It should be remembered that if the intake air does contain these ice crystals, carburetor heat might actually cause carburetor icing by melting the crystals and raising the moisture-laden air to the carburetor icing temperature range.
13. Prior to closed-throttle operation, such as for a descent, apply full heat and leave on throughout throttled sequence. Periodically, open throttle during extended closed throttle operation so that enough engine heat will be produced to prevent icing. Be prepared to remove carburetor heat if go-around is initiated.
14. Return control to "cold" position immediately after landing. If carburetor heat should be further required, observe ground precaution in No. 4 above.

In conclusion, many accidents concerning carburetor heat continue to happen each year. Many times it is not from the failure to recognize carburetor icing symptoms, but rather the neglect of a simple control knob on the the panel. Fundamentally, it doesn't matter whether one recognizes carburetor icing right away. Every pilot should have set procedures for engine failure. Among the first items to be done is to pull carburetor heat or change to the alternate air source. If this is done automatically, the problem may be solved automatically. Too many times pilots, or people in general, tend to believe the worst and overlook the obvious. In other words, if the engine loses power and runs rough, many pilots think of all sorts of major remedies instead of pulling a simple carburetor heat knob.

Taking off on a cross country with a student one afternoon, the Cherokee 140 began to shake vigorously. Over a large city and only 1,000 feet above the ground was no place to lose an engine. My training for emergencies was engrained. Mixture rich, carburetor heat warm, fuel pump on, and switch tanks—that was the checklist. Now, check the magnetos. Taking the plane from the student, my

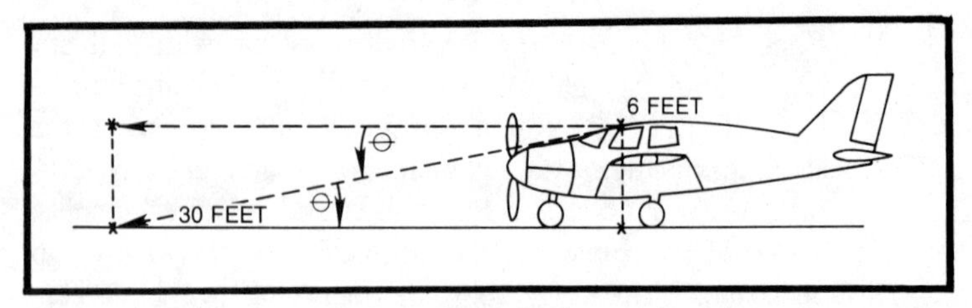

Fig. 4-7. Knowing the cockpit cutoff angle of your plane helps in estimating inflight visibility. Then a pilot can make a decision to continue, file IFR or make a 180.

fingers flew from one item to the next as I turned back toward the airport. The landing was normal and the airplane shook all the way to the ramp. The cause? Well, it was a broken valve.

This emergency was fairly major. The lesson, however, is to check the obvious things first. It could easily have been carburetor icing and the trip could have been continued. There's no use in becoming rattled over a little ice that can be melted with cockpit controls.

How to Determine Cockpit Cutoff Angle and Estimate Inflight Visibility

What does the ability to estimate inflight visibility have to do with cold weather flying? It has to do with continuing VFR or filing IFR, that's what. Many accidents every year are explained by the phrase "continued VFR into adverse weather conditions." If a method could be derived for estimating inflight visibility quite accurately, then that information could be used to stay within basic VFR weather conditions such as 1,000-foot ceilings and three-mile visibility. If a pilot could tell when the visibility was dropping below three miles, perhaps the decision to file IFR or turn around would be more clear-cut (Fig. 4-7).

The process begins on the ground and can be part of the fall preparation for winter. Adjust the aircraft attitude as closely as possible to the normal cruise pitch attitude. Get in the pilot's seat and adjust it to the same position you would use in flight. Also, it is important to assume your normal posture. Put a mark on the side window next to you at a position even with your eye. Now, look over the nose of the airplane (straight down the cowling) at a point where an object is just visible. This is the cockpit cutoff angle.

Now, you should climb outside and measure the height from the ground to your eye level mark (example six feet). From a point directly below your eye mark, measure the distance along the surface to the object you spotted from the cockpit (example 30 feet).

The tangent value can be calculated by simple division:.

$$\frac{6}{30} = .20 = \text{tangent value}$$

Check Table 4-4 for the tangent value of your airplane and read the visibility in the far right column. That value can be related to miles quite easily and then, when the weather starts coming down you will know when to quit or do something different.

Detection and Prevention of Carbon Monoxide in Aircraft

Carbon monoxide is the by-product of incomplete combustion of carbonaceous material. It is found in varying amounts in the exhaust fumes and smoke produced from burning fuel and lubricants. Carbon monoxide itself is odorless, colorless and tasteless, but it is usually associated with other gases that can be detected by sight or smell.

Hemoglobin is the oxygen carrying agent in the blood. The affinity of hemoglobin for carbon monoxide is higher than that for oxygen. Whenever carbon monoxide is taken into the lungs, it will stick to the homoglobin first and result in oxygen starvation. Oxygen starvation affects the brain directly in its functions to reason and make decisions. Exposure to small amounts of carbon monoxide will reduce a pilot's ability to operate an airplane safely. It can be considered that prolonged exposures to small amounts of this toxic gas are as dangerous as a short exposure to a relatively high concentration.

Table 4-4. Estimating Inflight Visibility at 1,000 AGL Based On Calculated Tangent Values.

TANGENT VALUE	ANGLE (DEGREES)	APPROX. VISIBILITY AT 1,000′ AGL. (FEET)
.052	3	19,200
.070	4	14,280
.087	5	11,500
.105	6	9,530
.123	7	8,130
.141	8	7,090
.158	9	6,330
.176	10	5,750
.194	11	5,150
.213	12	4,710
.231	13	4,320
.249	14	4,010
.268	15	3,730
.287	16	3,480
.306	17	3,270
.325	18	3,070
.344	19	2,910
.364	20	2,750

As we all know, the concentration of oxygen in the air decreases with altitude. If carbon monoxide is also in the air at a high altitude, the body will not get enough oxygen and the situation becomes dangerous. Thus, it is appropriate to say that susceptibility to carbon monoxide poisoning is greater the higher we fly. Inhalation of tobacco smoke also introduces CO into the body in significant quantities.

Many light aircraft cabins are warmed by air that has been circulated around engine exhaust pipes. A defect in the exhaust pipes or cabin heating system is the most likely source of introduction of carbon monoxide into the cabin atmosphere. Winter is the most dangerous time of the year because that is when we are most likely to use the cabin heating system and keep windows or vents closed. The cabin heating system is not the only area that can introduce carbon monoxide into the cabin, however. Holes in the firewall and perforations in the fairings around exhausts can be sources as well.

There are some early symptoms that will help pilots recognize the onset of carbon monoxide poisoning. These are feelings of sluggishness, being too warm and tightness across the forehead. The early symptoms may be followed by more intense feelings such as headache, throbbing or pressure in the temples and ringing in the ears, if the condition persists. These, in turn, are followed by severe headache, a general weakness, dizziness and gradual dimming of vision. If some sort of action is not taken to get fresh air, sufficient CO in the body results in loss of muscle power, vomiting and convulsions and then coma. The end is near as the pulse weakens and the respiratory rate fades.

The measures to take if a pilot and/or his passengers show any of these symptoms are simple and straightforward. Immediately, the cabin heater should be shut off and any other openings that might be responsible for exhaust fumes should also be closed. Usually exhaust fumes are smelled before the symptoms manifest themselves.

Other things that can be done are to open a fresh air source immediately, avoid smoking and inhale 100 percent oxygen if it is available. If airborne, land at the very first opportunity and ensure that all CO poisoning symptoms are gone. Before taking off again, find the area that is allowing the exhaust gases into the cabin area and have it repaired.

Early in the book we discussed the way to get ready for winter in the fall. We discussed inspecting the exhaust system and said that fall is the best time to do it, before the weather gets too cold. Pilots may not be aware that cracks and holes may develop over a short period. Because this is true, it may not be a bad idea to have the exhaust system checked every 25 hours or so. Carbon monoxide in the cabin has been traced to worn or defective exhaust stack slip joints, cracks and holes, openings in the engine firewall, "blowby" at the engine breather, defective gaskets in the mufflers, and inadequate sealing of fairings around strut fittings on the fuselage. As you can see, there are many places that could wear out and cause problems; therefore, frequent checks are desirable.

There are other ways to experience CO poisoning. You might be surprised to find that following a jet on takeoff can do it. Taking a position downwind from a jet that is holding on the taxiway for takeoff is a good way to catch a whiff of CO. This is easily corrected by parking upwind of these fellows while you wait your turn for the runway.

Another good idea is to have a carbon monoxide test conducted on the ground and in-flight to determine the extent of any possible contamination. These tests should be run with cabin heat both "on" and "off."

Many mechanics have carbon monoxide test devices. There are two types. One is a syringe that draws air into a transparent tube containing material that is sensitive to CO. This material changes color and is compared to a sample tube which remains unchanged. In this way an accurate measurement can be taken.

The other type of CO indicator consists of a porous plastic disc, about the size of a dime, mounted in a solid plastic plate about two inches square and one-tenth of an inch thick. The porous plastic will change color if it comes in contact with carbon monoxide. The color can then be compared to an instruction card and a rating of "safe," "marginal," or "dangerous" can be determined. This device is not as accurate as the syringe method, but it is adequate and much less expensive.

There are several CO indicators on the market that mount on the panel of a plane. Some of these are not accurate or have never proven their worth. The FAA has only tested one such device, so be sure that the one you obtain is the tested version or comes with a guarantee that it works.

Mountain Flying in the Winter

There is enough information on the techniques of mountain flying to fill a book. The purpose of this section is not to totally describe mountain flying, but to touch on the areas that apply to winter.

Having flown a mail route from Spokane, WA, to Lewiston, ID, to Boise, ID, and back five nights a week, I learned many things. The route from Lewiston to Boise was the tough part of the route. The minimum en route altitude for instrument operations was 12,000 feet. The Victor airway took me down the Snake River to Hell's Canyon, reputed to be the deepest canyon in North America. It is about 7,600 feet deep. From the canyon the route flew down the Salmon River of No Return towards Boise. The simple fact is that the route was one of the toughest mail routes in the Continental United States. The first lesson I learned was that in the high country there are few weeks when it is not winter. Snow covered the peaks all year around.

Another thing that became very apparent was that night flying in the mountains added another dimension to the risks involved. Flying at night is not inherently risky; crashing in the mountains at night is. Though my opinion may seem ultra-conservative, several engine failures made it apparent that a forced landing in rugged country would probably not be survived. Thus, my first rule of mountain flying, winter or summer, is *don't fly at night unless it is your job.* If a night flight is a must, then file IFR unless you are *extremely* familiar with the area and have flown the trip many times before. A VFR flight at night is more risky, especially if the night is moonless or marginal VFR. In that case, the flight should not be made because marginal weather operation is doubly dangerous in the winter. Alternate courses of action, or changes in destination, are always limited because of the terrain. Mix winter weather with those normal limitations and the alternates become extremely scarce.

The people of the West have a saying: "The mountains make their own weather." That is plainly true. Take Denver, for example. Many times the mountains to the west are filling up with snow and the sun prevails in the city. The significance is more important than this simple observation. Distances in the mountainous West can be very vast and lightly populated. The point is that there are few weather reporting stations for the areas. Couple this with the fact that the weather may be doing something entirely different on the other side of the ridge from where the observation was taken, and weather information becomes very sketchy. Also, this sketchy collection of data is fed into the big picture. Assuming that the mountains can and will disrupt normal weather trends calls for more than the average alternate plans of action. Flatlanders should talk to the locals to obtain an idea of the weather situations that frequently occur.

As unfortunate as it is, we must still depend on weather trends and forecasts to plan our flights. The best advice is to pay extreme attention to all possible alternate plans when gathering information during the preflight briefing. It is a good idea to set intermediate goals for a trip. For instance, plan to go to point X if the weather is too bad to continue to point C when arriving at point B. It is most important for a pilot to recognize his own limitations and abilities, keep an open mind toward changing plans and refrain from continuing ahead at all cost. The price may be higher than a pilot would want to pay.

A maneuver that is sure to get a pilot in over his head (literally!) is to fly into snow or rain showers that obscure the terrain. The 180° turn will buy some time and allow a pilot to look and consider an alternate plan of action. On this count, there is a method of navigating that is not as risky as blind faith. On occasion, I have flown over a highway through a pass during a snow shower. This occurred in the state of Nevada and it worked because I knew the terrain didn't rise very high above the road in the area I was transiting. Following a highway in this fashion could prove to be dangerous if a pilot did not know his precise location on the Sectional. For example, the highway could disappear into the snow shower and then into a tunnel. Good judgement, knowledge of the aircraft's precise location, and intensity of the snow shower are the factors that must be considered before sticking the nose of the plane into the slop. If the visibility goes below a mile in the snow shower, then this sort of operation is much too risky.

Mechanical operation in the mountains involves the regular cautions that we have discussed for all winter flying. However, since altitudes flown in the mountains are generally high, the temperatures aloft may be much lower than the temperatures out on the plains at lower altitudes. When letting down to our destination, it may be difficult to keep the engine warm enough. Quick engine cooling can be harmful and shorten engine life. As a result it may be necessary to let down with power. This will affect one of two things: It will increase the ground speed or the rate of descent must be increased. The let down should begin as early as the terrain will permit in order to keep the rate of descent at a tolerable level. Unfortunately, there are ridges that surround some mountain destinations. In this case, a slow descending spiral will ease the plane to the ground, enabling the engine to remain warm and the ears to adjust.

Descents are the toughest to make in airplanes equipped with fixed-pitch propellers. When the nose is lowered the airspeed will increase as will the engine rpm. To keep the engine within rpm limits, the throttle must be retarded. A compromise must be struck between keeping the engine warm and engine rpm tolerable.

The altitude engine is the easiest to handle in descents. The manifold pressure can be brought back just enough to start the plane downhill and left in position. The throttle will still need to be adjusted every 1,000 feet or so. Turbocharged engines with automatic waste gates are handy because the manifold pressure remains where it was set. Turbocharged engines typically run a little hotter and one eye should be kept on the cylinder head temperature. Keep the gauge in the green.

White-out conditions are possibilities in the winter. These conditions can be brought on by several different weather phenomena, but haze and fog are the principal causes. Snow-covered terrain tends to blend into the bad visibility as if it isn't there. Falling snow and/or blowing snow restricts the visibility to white-out conditions as well. A pilot, IFR rated or not, may find himself flying instruments due to a complete loss of visual contact.

One of the best preflight plans a pilot can make is another appointment. Always tell a business appointment that you intend to fly. If conditions "rain you out," so to speak, then tell the contact that you will make other arrangements to arrive at a later date. This takes all the pressure off. If the flight can't be made, take an alternate form of transportation.

Remember, the pilot alone is responsible for the "go, no-go" decision. My flying attitude has always been to go, or at least start, unless the weather is the worst at the departure point. Then careful evaluation of the weather as the flight progresses is necessary and the points can be picked where to deviate or turn back. However, my philosophy is one of a professional nature and may not be the best for a neophyte pilot. In that case, a good position to take is—"If in doubt, don't!"

In addition to all normal winter precautions, a pilot may be faced with some other problems in operating an airplane in the mountains. These problems are the normal, everyday troubles with which experienced mountain pilots must deal. It probably is not a far-fetched assumption that if you are reading this book you might want to plan, or are planning a trip to the mountains in winter. Addressing that assumption, let's go over landing and takeoff techniques. We'll begin with landing since a landing in the mountains must be made before a takeoff can be made.

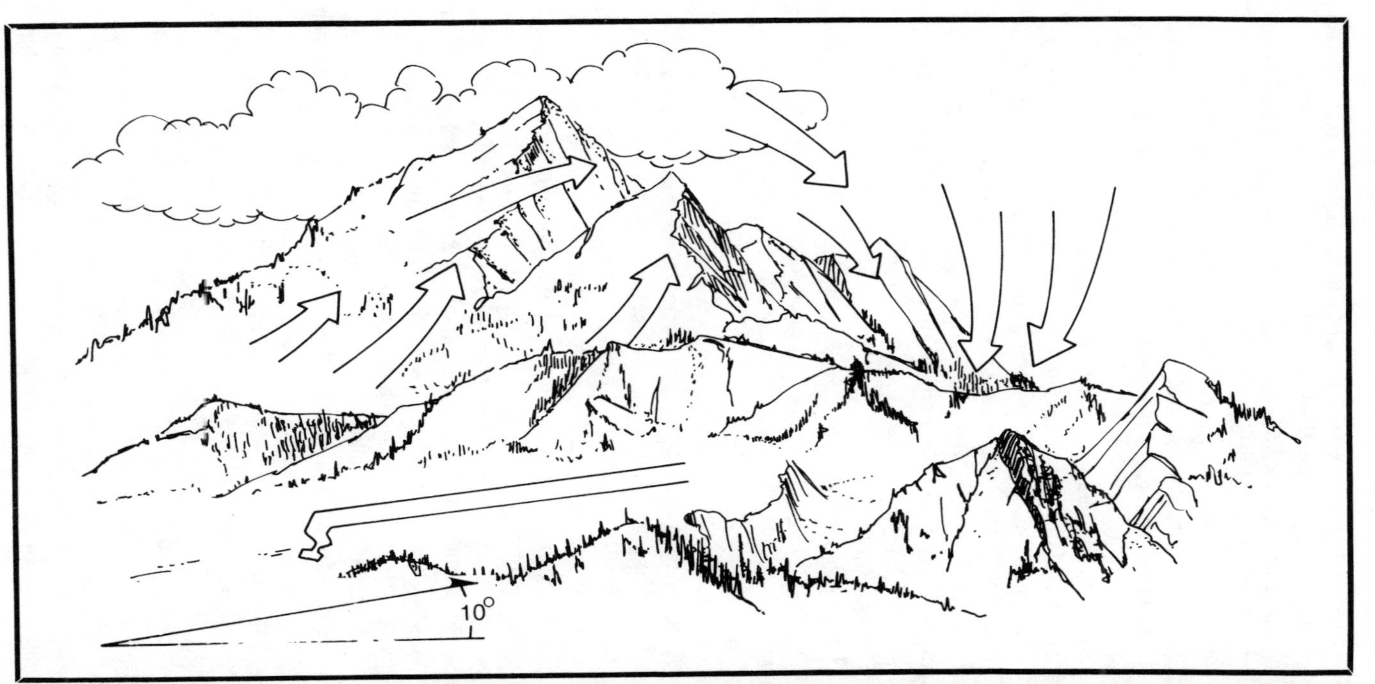

Fig. 4-8. Before landing in the mountains, check the terrain and how it will affect the way the wind will blow. As you see in this drawing, there is no other way to align the runway in the valley.

As we discussed earlier, it is not unusual for a pilot to find himself arriving at the destination with an excess of altitude. This may be due to the natural desire of wanting as much altitude in the bag for safety as possible. The location of the landing area may be low in relation to the surrounding terrain. Also, as pilots, we tend to fly higher to extend the distance we can see ahead and sometimes we simply fail to recognize our destinations far enough in advance (Fig. 4-8).

Landings at unfamiliar airports usually need to be planned. Flying over the airport to observe the wind direction and layout are necessary anywhere. An approach into a mountain airport should include a look for obstructions on the final approach and how the terrain will affect wind flow patterns closer to the surface. The descent can then be planned in such a way thathe aircraft is positioned in the best place to handle turbulence, rapid drift, wind shear or downdraft.

These rules sometimes change in mountain flying. Full rectangular patterns may be out of the question at some high terrain airfields. It may be necessary to fly down a canyon on the left side or sunny side while making a controlled descent, then make a quick 180° turn to line up on final.

Another typical problem in the mountains is the one-way airport. These airports have approaches, for example, from the south and takeoffs are to the south. The reasons are not usually for noise abatement, but because of rock walls or a sloped runway.

Sloped runways can cause a couple of unusual difficulties. Experience in these situations is an important asset; but, if a pilot doesn't have the experience, proper preparation will help. Many times the Airman's Information Manual, Part II, will state whether the airport is one-way and has a sloped runway. Also, the AIM often names obstruction to each available runway.

The difficulties we alluded to arise at any runway that is sloped. The biggest problem comes with the landing approach. An approach to a runway sloping uphill, away from the aircraft, results in the illusion that the approach is too steep. In this type of situation it becomes easy to shallow the approach to a dangerously low altitude without realizing it. The exact opposite occurs if the runway is sloped down and away from the plane. The approach in a normal position has the illusion of being too flat and the tendency is to come in too high and with too much speed. That is exactly what a pilot does not want to do, because on a downslope runway it is much harder to dissipate speed and much more braking is necessary. Remember, each runway is only so long and if the approach is too hot just prior to the threshold, you should go around if terrain permits.

The danger of an upsloped runway is landing short. If the pilot is aware that the runway slopes uphill in the direction of landing, he should carry a little extra power to keep from becoming too low and slow. The fact that the touch-down will be uphill means speed will dissipate more quickly.

Visual Approach Slope Indicators (VASI) have been installed at many mountain airports to eliminate these illusions. One must realize that at isolated mountain strips (out where the fishing is good) these electronic wonders won't be available.

The effect of a sloped runway is complicated in the winter, as I witnessed one snowy morning. The weather was improving at the destination. All night, light snow had been falling turning the mountain valleys into Christmas card scenes. The peaks were all hidden in the low clouds, but the visibility was generally three to five miles except in the snow showers. There was no way to fly VFR that morning, so we saddled up and took off IFR.

The flight progressed routinely. We changed frequencies, chatted with the controllers and finally were cleared for the approach. The VOR approach had low enough minimums for the weather, so there was no concern that I would not see the airport.

Little did I know that while I was completing the procedure turn, the snow plow was finishing his last swath on the runway. As he turned around to head for the barn, he clipped the VASI installation with the blade. The VASI was out.

Like every good instrument pilot should, we ascended to the MDA as quickly as possible. We broke out early and, of course, the MDA had us low. The usual technique is to fly along level until intercepting the glide slope on the VASI and then follow it down. The new snow blanketed everything in sight. The world looked white and the runway was out there somewhere straight ahead.

Finally, we spotted the hangars and could make out the runway. It was snowpacked and no sand had been put out. My depth perception was nil. Looking for the VASI we found two boxes protruding out of the snow and that was it. I set up for the approach normally. As we got close to the runway I noticed the need for more power—then a lot more power. We touched down in the first ten feet of plowed runway and rolled quickly to a stop. Looking straight ahead I could see plainly that the runway was sloping up. Though I often flew into the airport, I always used the VASI and never really thought about the slope. The bad depth perception, coupled with the illusion created by the sloping runway—not to mention starting low from the MDA in the first place—was enough to threaten to ruin a good pilot and his family.

While we were in the lobby, the snow plow driver walked in. He was not aware that he'd hit the VASI yet, but he had seen our approach.

"Say, are you the pilot that just flew in?" he asked.

"Yes, I am."

"Boy, you're pretty good. Your wheels just skimmed the snow the last 25 yards to the runway. Ain't never seen anything like that. Kinda like Jesus walking on the water." He was impressed. He turned away to talk to his boss.

I thought to myself, "Like Jesus walking on the water, huh?" I could have killed myself out there.

The truth is, believe it or not, I've never had that problem again (knock on wood). The lesson I learned that day is always carried with me.

Another factor that relates to the new pilot venturing into the mountains is *density altitude*. In the wintertime we don't have hot temperatures, extremely long takeoff rolls and that sort of thing, but density altitude doesn't just mean summer problems. In winter at high altitudes, the effectiveness of the wings, propeller and air impact system are diminished just as they are in summer.

As you have probably noticed and have been taught, indicated airspeed diminishes with altitude. True airspeed generally increases with altitude to a point where the engine works at optimum standards. What this has to do with mountain flying is the use of proper approach speeds; simply put, an airplane that stalls at 60 knots at sea level will also stall at 60 knots at 9000 feet. However, many pilots falsely assume that airspeeds should be lower on approach in the mountains. This is definitely a hazardous idea. An airplane landing at a 9,000-foot altitude will have a much greater groundspeed (in no wind) over the threshold at normal approach speed than it will at sea level. The reason is simply because the air is less dense and the airplane must move through that air more quickly in order to produce the same lifting effect as at sea level.

The approach speeds for the airplane you fly should be kept at the usual values! Those values are usually 1.3 Vso or 1.3 times the stalling speed of the aircraft in landing configuration. In wintry winds another good idea is to add a gust factor of simply 50 percent of the gust velocity. If the wind is gusting 18 knots in excess of the steady velocity, then nine knots should be the gust factor used.

Many light aircraft have little or no excess horsepower left when landing at high airports. Though the FAA recommends full flaps for most landings, they may be a hazard at altitudes as low as 4,000 to 5,000 feet. The power to overcome the rate of descent may not be available. Therefore, for a full flap landing, wait until the field is made before lowering the flaps to the full setting.

Of course, it goes without saying that winter landing techniques will be needed in the mountains as well as elsewhere. Don't leave your good sense at home. The mark of an excellent pilot is if he can apply what experience he does have to new and interesting situations (Fig. 4-9).

Pilots that are new to the mountain environment can expect to be under a physical and mental strain. These stresses can be strong enough that they may steadily erode a pilot's capacities for thought and action. By the end of a flight, the ability to cope with the approach and landing will be at its minimum; but frankly, that is when the

Fig. 4-9. We have to live with crosswinds in the mountains. As you see in this drawing, there is not another way to align the runway in the valley.

demand is greatest. In order to compensate for those deficiencies and reduce the amount of stress on the pilot, a few good rules can be followed.

The trip should be carefully planned to transit the most favorable terrain possible. There's no reason to fly over the biggest mountains and deepest gorges. If you want to see them up close, backpack or take a jeep into backcountry. The trip is really more awe-inspiring and the stress of making a mistake is decreased.

Peace of mind can be directly related to the availability of survival equipment and supplies. If the trip must cross uninhabited areas, carry the necessary excess equipment. You'll feel a lot better for it. We'll talk about what to take in the last chapter.

One item always on a pilot's mind is the weather. Watch the trends on the TV weather for several days before beginning your trip. Another procedure I use and recommend is to check the weather every hour along the remainder of my route and at the destination. Trends can be spotted well ahead of time and the necessary deviations and amendments to plans can be handled without pressure. Worst of all is waiting until you can't see past the windshield to do something.

Call ahead to see what services are available. It may be possible to hangar the airplane and keep the snow off the hinges and wings. Ask if they have preheat and if the runways are plowed regularly. Las Vegas, New Mexico, is a 6,000-foot high area with frequent snow. Yet, the runway gets plowed after the town gets plowed—which may be several days after a storm. Knowing a fact like this ahead of time can save grief.

Even though fuel is available at the destination, carry an hour's reserve. As discussed many times already, the weather changes rapidly in the mountains and a snow shower might preclude a landing. With an hour's reserve, one can find another airport, fuel and wait for the weather at the original destination to clear up.

Upon arrival in your termination area, carefully evaluate the local terrain, obstructions and wind. Plan the approach in order that the situation will not become untenable. In the mountains, expect turbulence as you descend below the peaks. I always keep my descent speed at or below maneuvering or gust penetration speed. By expecting turbulence and already having the airplane configured to handle it in the best possible manner, it is easier to fly. The stress is less. When I was learning to fly in the mountains, the turbulence often scared me; hence, I came up with this procedure.

Many times on landing approaches I thought the bumps would

securely exceed the limits of the airplane. Sometimes, I would pull up and go around and try to stay clear of that area on the next attempt. The secret of flying mountain turbulence came with experience. Fundamentally, I had to learn how much the airplane could take and then how much of that I could take. The airplane invariably could take more than myself; so then the technique became one of slowing before arriving at points of expected turbulence. By flying at gust penetration speed the wing will stall before it will damage. Now, this may sound like a strange attitude to have, but if the airplane isn't stalling and is within air speed limits the flight or approach can be continued safely. Granted, it may not be continued comfortably, but those are two different things.

Upon touching down, continue to be wary of crosswinds, and winds disrupted by terrain. If the 90 degree crosswind component exceeds 2/10 times the stall speed (Vso) in landing configuration, it is usually too much for the average pilot or light aircraft.

To increase braking efficiency and all-around controllability on a slick surface, the flaps should be raised after touch-down.

Lastly, if any part of an approach is not to your liking, then go-around. Study the situation in your mind for a minute. Look over the approach again from the air. Possibly try a different approach, but always be careful.

Out of all the do's and don'ts there is one rule that doesn't come up too often. This one does not strictly apply to wintertime, but winter is the time of year when the advice is best heeded. Winds at higher altitudes are generally of a higher velocity until high in the stratosphere. In the winter the polar jet dips farther south and winds aloft are much higher than in summer. Since mountain flying requires operation at these higher altitudes, frequently between 10,000 to 16,000 feet, winds may be too strong for light aircraft. Basically, one should not venture out if the winds at one's proposed altitude will be in excess of 35 knots. The reason for this is that turbulence will be moderate to severe for many thousands of feet above the ridges. If the wind is to be a head wind, then more fuel than can be carried may be required to reach the destination. Also, once winds reach a certain level or velocity the accuracy of their predicted velocity recedes. What is predicted as a 35-knot wind may become a 50-knot wind easily. Fifty knots can be almost half of a worthy airplane's true air speed. Therefore, when winds forecast exceed 35 knots then it is better to sit it out on the ground. Go back to the lounge at the ski resort and watch the bunnies.

Once again, I should mention that all these hazards exist sum-

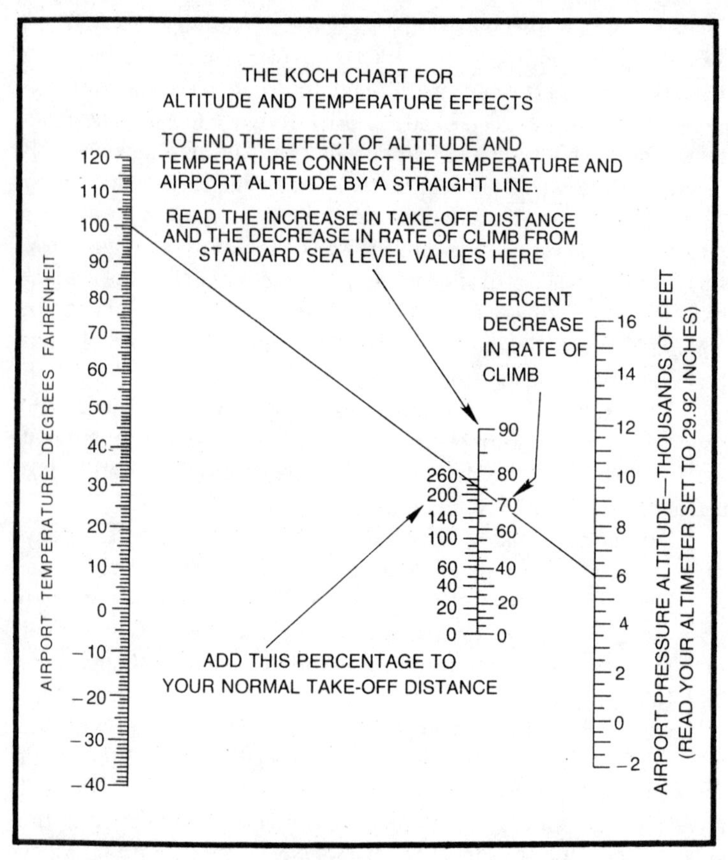

Fig. 4-10. Temperature can affect flight any time of the year. Work a few sample problems and see what the results are.

mer and winter. The reason we are examining them is that many pilots decide to take their first winter trips because of the lure of the mountain ski areas. Therefore, one should be armed with a little general knowledge on mountain flying. The main idea to remember is that winds in the winter are frequently stronger than their warm weather counterparts (Fig. 4-10).

All pilots have heard the grizzly stories of mountain downdrafts. These demons of the air can be avoided at least 95 percent of the time by using common sense. The air flowing across the mountains should be thought of as a stream, flowing across and around boulders, tumbling down drop offs, and the like. If a pilot maintains an awareness of the wind at all times, he will have the necessary information to make judgements. As the wind strikes a mountain situated in front of it, it will rise. On the back side, or lee side as it is called, the wind will descend. This is where the downdrafts are found. Therefore, it is best to fly on the windward side of a mountain. The updraft can be used to push you up and away from the terrain, whereas the downdraft will suck you down into the rocks.

The best thing, though, is not to get excited when caught in a downdraft. It will happen occasionally, so put any thought that it won't out of your mind. The good news is that downdrafts usually cease, leaving the airplane enough room to maneuver away from the terrain. Don't count on this fact, though, in extremely turbulent air.

If you do get caught in an extreme downdraft, apply full power immediately. Maintain best rate of climb speed. If there is plenty of room, turn away from the mountain. As the downdraft pulls you down, the terrain will also be diminishing, giving the best possible chance to outdistance the demon before contacting the earth below. At all costs, don't stall the plane. Remember, at higher altitudes the best rate of climb will be a slower airspeed than at low altitudes. This is because the efficiency of the wing is diminished by less dense air.

Out in the Midwest and Great Plains areas, the days sometimes get hazy. That, coupled with the snow-covered ground, causes the horizon to become hard to define. This can also happen in the mountains: however, horizons in the mountains are rare, especially when approaching a mountain valley airport. The common mistake made by inexperienced pilots in the mountains is to use the peaks as the horizon. The truth is that the horizon is closer to the

Fig. 4-11. Airports are typically deserted in the winter, so calling ahead is the wise thing to do.

base of the mountain than the peak. The mistake of using the summit as a horizon reference will result in a constant climb. An inadvertent stall could result and spoil the whole day.

When approaching a ridge, it is a good idea to use a 45 degree angle approach. If a dangerous downdraft is encountered, then only a 90 degree turn will get the airplane headed for lower terrain. If a straight-on approach is used, a 180 degree turn will be necessary and these take more time to execute.

These are a few of the major cautions in mountain flying. Very fewe flights are routine in high country flying. Expect to encounter the unexpected because everyone does. Follow good operating practices and maintain a sensible perspective of the necessity of the trip. "Get-home-itis" has killed many a pilot and innocent passenger. I usually try to plan trips so that I leave a day ahead of time, giving me a 24-hour pad. Weather generally is not raunchy for more than 24 hours at a time. When I do arrive back home a day early, I can usually put the time to good use around the house. The reason for doing this is to take the pressure off and let me enjoy flying to the maximum. If a flight has to be completed within a certain rigid time limit, don't fly. Take an alternate form of transportation, such as the airlines. Their completion rate is tremendous.

What to Expect When You Get There

The worst thing a pilot can do is to take off without ever checking the Airman's Information Manual. To arrive at a desolate, ramshackle, old airport in the dead of winter produces one of the lowest feelings a pilot can have. Imagine two rusty old T-hangars rattling in the wind and a small office over on one corner of the ramp with the windows busted out. The telephone on the outside wall was installed by Alexander Graham Bell and probably hasn't worked since they retired Hattie the operator. It must be five miles to town and the road, coming in, looks like it is only fit for snowmobiles. You wonder, "Why didn't I check this place out before we left?" (See Fig. 4-11).

It's the truth, standing in below freezing temperatures at a location like this is nowhere to be. To quote a famous TV commercial, "What *will* you do?"

Now that we have realized the bleakest, dreariest possible misfortune, it should be easy to plan for the future. The Airman's Information Manual is handy for finding out about strange airports. Another good source, and possibly a more up-to-date one, is the Aircraft Owners and Pilots Association's (AOPA) Airport Directory. Between these two publications a pilot should be able to find the type of fuel available. This the most important question to be answered. (By type, I mean 80/87 or 100/130 octane or the newer 100 LL, as well as the brand.)

The next most important question is hours of service. It is far better to arrive when someone is attending the airport, than otherwise. Valuable information can be obtained from people familiar with the area. Also, it is better to be told where to put the flying machine for best protection than to make a guess. Inside phones will usually be accessible for a call for transportation if necessary. Stay away from airports that have no set hours, at least for prolonged stays. The services available at these wayposts are generally not acceptable for hard-core winter operation. They probably won't have things like preheaters, de-icing equipment, fuel additives etc.

If a stay is to be prolonged in an area, transportation is a problem to be considered. Many general aviation airports have rental car agencies and these are sometimes listed in the airport directory. Sometimes, the surrounding area is so small in population that only taxis or courtesy cars are available. Plan ahead and make reservations prior to the departure date.

When not staying in an area where friends are near, lodging is a necessary consideration. The AOPA Airport Directory usually lists two or three lodging establishments that are near the airport. Most people are like me, though, I imagine. Unless I am familiar with a spot or one has been recommended, I tend to stay with the brand name hostelries. The overall condition of an establishment can then be judged, though if doesn't always hold true.

The availability of tiedown space or hangar space cannot usually be determined from these airport directories. However, a telephone number is usually given for one or several FBO's on the field. An advance phone call can determine the availability of hangar space and preheat equipment. Don't fly to an airport that doesn't at least have tiedowns available. In the summer months it may be possible to bring your own "screw-in" tiedown anchors. Winter freezes the ground, however, and putting a tiedown anchor into the ground might require an air hammer.

If no hangar space is available, a good investment is wing covers. These items make preflight in the winter easier to deal with. Just unsnap the covers and remove all the frost, snow and ice—at least the lion's share of it.

Finally, an item to check is the type of instrument approach facilities located on the field. A quick look through your Jepp charts will tell you that, or you can check Part III of the Airmans Information Manual. The better the approach facilities, the worse weather an IFR rated pilot can hazard in making the trip.

This type of preparation can start days or weeks in advance. The pleasure trip will be a pleasure and the business trip will get more business done when the pilot isn't standing out in the driving snow listening to the two old rusty hangars creak.

There are many considerations and operational techniques for cold weather flying. The best advice is to always be thinking ahead of the airplane. That kind of thinking will keep the wheels on the bottom and the pilot on top. Soon, the operational considerations will become familiar and the techniques polished. In the next chapter we'll discuss the airborne aspect of winter.

Chapter 5
Flying the Weather

An airplane is a mighty ungainly creature on the ground with its spindly legs holding up a massive torso and wings like planks. The front of the plane is smooth and streamlined, yet the propeller looks like a beanie cap that has slid down on its nose. Tied down it looks like a giant pterodactyl the Japanese capture in a horror movie. But in flight—ah, what beauty.

Truly, one of the most beautiful sights to behold is that of another airplane flying nearby. The air currents bubbling up causing it to bob gently give it life and majesty. The airplane is one of Man's greatest creations, spanning vast distances in minutes and carrying us in comfort and style.

The problem with airplanes is that they are affected by weather directly. Probably no other activity is as often touched by changing air masses as is flying. Surely, the ships at sea are affected by storms. The farmer depends on rain to raise his crops and the hail sometimes beats them down. But the airplane—the airplane is either blessed by the wind, washed by the rain, kissed by the sun, cursed by the snow or bedraggled by the ice everytime it leaves the bonds of earth. Nature constantly issues a challenge to man's creation and we as pilots must meet that challenge in order to survive until the next flight.

Winter, in all of its silent beauty and howling fury, is one of the toughest challenges for the flyer to meet. A pilot doesn't need to be instrument rated to fly in the winter, but it helps. The VFR pilot can make trips just as well, though not as often. Thus, every flight should begin with the question IFR or VFR?

The distant voice on the other end of the phone was describing the position of the low nearest my route of flight. Mundanely, he described the drape of the cold front across our flight routing. The reporting stations were all reporting at least light snow falling and some had moderate pockets of precipitation. The visibilities were all marginal, if not IFR already, due to some fog and some blowing snow. There were no icing pilot reports. The choice this day was easy—file IFR.

Some days the choice is not that easy. There are days when VFR operation is preferable to IFR. Those days relate to the type of equipment we are flying and the forecast conditions.

One thing that should be said right up front, before going any further, is that planes without any de-icing equipment can be flown in snow. The Cessna 172, which is America's most popular type of air transportation, has no de-icing equipment except a heated pitot tube. (At least I have never seen one with any more equipment than that.) On many days throughout the winter, in any part of the Snow Belt, that airplane can be safely operated in snow.

As long as the snow is not wet or nearly rain it will not stick to the airframe. One reason that this is true is because the snow is solid. Like the ice crystals in the cirrus clouds, the snow flakes tend to bounce off the airplane. Another reason the snow flakes do not cause airframe icing is that they never touch the plane. That's right—most snowflakes tend to follow the slip-stream around the aircraft surfaces and then continue their downward plunge toward earth. What most pilots don't know is that a wave is set up in front of each of the plane's surfaces such as the wing, cowling or empennage. This wave, though it doesn't extend more than two or three inches ahead of the surface, is sufficient to lift the snow flakes into the slip-stream and, thus, they avoid the airframe.

When the snow is wet or heavy the wave ahead of the surfaces is not strong enough to lift the precipitation particles into the slip-stream; nor is the slip-stream of sufficient velocity to carry it. Thus, rain or wet snow does reach the airframe.

The importance of having this knowledge should be clear. Just because it is snowing, it may not be reason enough to cancel a proposed VFR flight. VFR flights *can* be made in snowy conditions. Snow isn't the limiting factor ordinarily. *Intensity* of the snow shower is, because it relates to visibility and ceiling.

Why Fly VFR at All?

There are several answers to this question. VFR is necessary to those not instrument rated, obviously. We mentioned the limitations of equipment earlier. Many general aviation planes do not have de-icing equipment and many cannot fly above the common icing altitudes. This includes especially those single-engine airplanes flown in the private sector of aviation.

Flying VFR, then, may be the only way to complete a trip on some days, or the trip might necessarily need to be flown *quasi-VFR*. This term, which I coin here, actually pertains to an IFR flight where every concerted action is taken to stay clear of clouds in order to avoid ice. Essentially, the only reason the flight is filed IFR is that visibilities are expected to be frequently below three miles in snow showers. Hence, quasi-VFR is staying VFR as much as possible, yet having the protection of radar coverage to help in avoiding others.

Let's look at a typical day when VFR would be preferable to IFR. Our intended route of flight is a 300 nautical mile course through snow country. The preflight briefing lists ceilings along our route of flight to be in the neighborhood of 3,500 to 4,000 feet. We are warned of scattered snow showers in the forecast and the current radar report confirms that they are out there. When we ask the briefer if there are any Airmets or Sigmets he responds, to our dismay, with an Airmet calling for moderate icing in clouds in precipitation. The freezing level is at the surface and the winds aloft forecast indicates that there are no non-freezing temperatures above. This situation is very often the case in late January and February after the atmosphere has had time to cool down significantly at all levels.

The briefer has painted us a very good picture of what to expect. If we fly higher than 3,500 to 4,000 feet, we will need a clearance because there are clouds. At those altitudes we can also expect, with assurance, that there will be icing. If the plane we are flying is not approved for flight into forecast or known icing conditions, we must remain on the ground or out of the clouds. Therefore, our only alternative is to commence the flight VFR. Actually this may be the only legal choice, but it is also the safest choice. Staying out of the ice altogether beats fighting it even if one has the equipment.

Many times during the winter the snow showers are scattered. This means they can be circumnavigated. On a day like the one in the example, some circumnavigation is necessary. Assuming that we stay VFR throughout the flight, we must stay in areas that have visibilities more than three miles. Going around snow showers is like going around thunderstorms—we must first ascertain the direction of the snow shower movement. More often than not, they move downwind from our winds aloft forecast. If the winds are predicted to be northwesterly, the showers will move southeasterly. A transition to the northwest side is appropriate. In VFR conditions I just eyeball it and go; but, for the fledgling, a more precise method might work better.

For example, let's assume that the magnetic course we are tracking is 030°. As we approach halfway on our trip the first snow shower sits directly ahead. From the winds aloft we know that these showers are moving southeasterly and the shaft of snow from the cloud is slanting in that direction. It is necessary, then, to go on the upwind side. Looking out the window we can estimate that a 30° correction to the left will take us clear of the snow on the northwest side. Turn the aircraft 30° to the left and note the heading. It will be 360, or north. Now check the time and write it down as we will need it later to get back on course. Now we hold 360° until we have cleared the shower. Some small deviations may be needed, but basically we should be heading due north most of the time. Once we are sure that we have cleared the snow shower area and a turn back to the right towards our course is clear, we check our time (Fig. 5-1).

Let's imagine that it took 13 minutes to clear the shower. Now, we must turn back to the right to get on course. The rule of thumb I have used in the past is to double the original deviation. Hence, if we originally left course with a 30° deviation to the left we should now turn right 60°. Now our heading will be 060 degrees and we have set up a 30° intercept with our original magnetic course.

After completing our right turn to 060, we time ourselves in that direction for approximately 13 minutes. Be sure to look at the Sectional or WAC chart often as the 13-minute mark approaches. The correction to the right in this example will take less than 13 minutes to intercept the course because of the northwesterly wind. The 13-minute figure will be more accurate, though, than a figure from off the wall. This method is much better than eyeballing and, in unfamiliar territory, will keep a pilot from becoming lost.

Life Cycle of a Frontal Wave

Winter is so much easier to fly if we have an understanding of the big picture. The big picture is knowing where the low pressure areas lie and how the fronts are draped. By knowing the characteristics of fronts we can, to some extent, predict the weather ourselves. Often, flying along, watching the weather out the windscreen is the

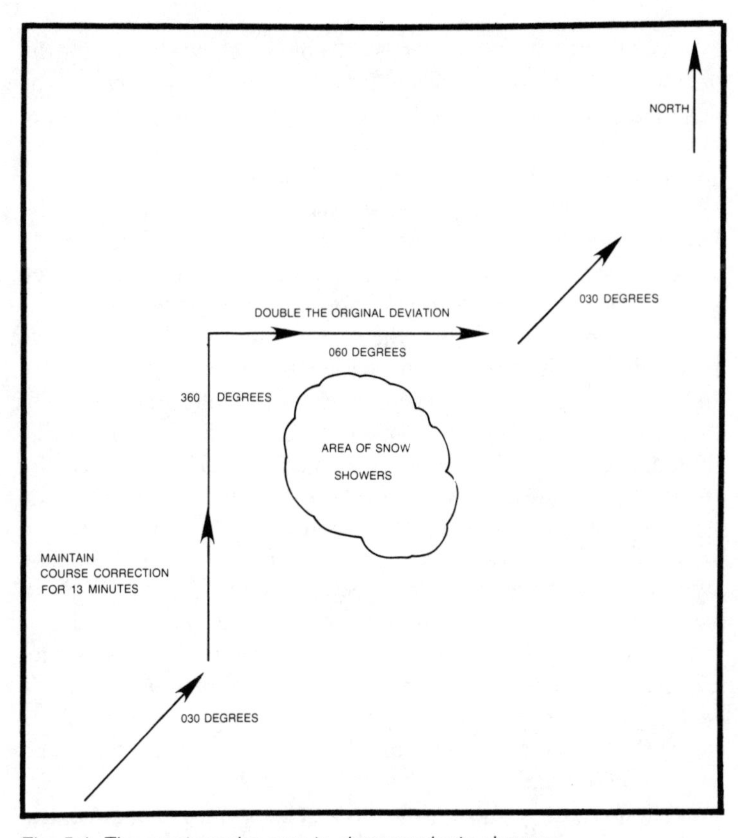

Fig. 5-1. The most precise way to circumnavigate showers.

best place to observe the current trend. It is so much better than the ground observer's position that a pilot can often forecast a change hours ahead of the Weather Service. This is chiefly because the airplane is getting a cross section of the weather and the people on the ground use only their observations and those of others miles away. It is on this point that the greatest incentive lies for checking the big picture.

When we check the weather from a Flight Service briefer he usually gives us observations from various points such as, "Syracuse 1,500 overcast, two miles and fog. Buffalo is reporting measured 500 overcast, one mile, light snow and fog." When this is the only information we get, then we cheat ourselves because the weather between Syracuse and Buffalo is left to the imagination. Is the weather at Buffalo low due to lake effect off of Lake Erie or is there a front of some sort in the area?

As a matter of interest, or as a hobby, I continually monitor the weather programs on television. I am likely to fly on a daily basis as an airline pilot, so it fits that I should know where the systems are and what they will do next. It is to this end that the life cycle of a frontal wave should be important to all pilots.

In the next few paragraphs, we will be referring to Fig. 5-2, the illustration on the life cycle of a frontal wave. The pointed barbs represent the direction of movement of the cold air and the rounded spades represent the direction of movement of the warm air.

In Fig. 5-2A the winds of the cold air and warm air masses are blowing in opposite directions but parallel to the developing front. As the air masses grow more in contrast to one another, the identity of cold and warm air becomes apparent. The barometric pressures differ across the front. This process is called *frontogenesis.* As the pressure differential between the two air masses increases, the wind increases. This sets up a ripple or the beginning of a wave. The beginning of the wave form is shown at B where it is becoming apparent already that this will be the main area of low pressure. In reality, a pivotal point for the fronts to move around is forming.

By the time the frontal wave has reached the adolescent stage, shown at C, the cyclonic circulation is well established. At this time a definite low pressure system can be depicted on the surface analysis maps. (Notice how the sequence of pictures actually looks like a swell in the ocean.)

The fronts have now grown to reach separate identities. The warm front is being circulated northward and the cold front is being pushed southward. The area to the south of both fronts (consider the top of each picture to be north) is called the warm sector. This area is where the moisture is the greatest. In the United States, the moisture is generally pumped up from the Gulf of Mexico into the central part of the country. The east coast often gets a mixture of Gulf and Atlantic moisture while the west coast derives its moisture from the Pacific Ocean.

The moisture is being lifted as it approaches both fronts. The lifting is more defined or extreme over the cold front, therefore, the cold front produces more inclement or violent weather than does the warm front. The moisture that is lifted must be eventually precipitated; so, as pilots, we must expect precipitation as we near a front. The type of precipitation depends on the temperature and the type of front we are flying through.

In D, the cyclone has grown to adult strength. The surface winds are sufficient to cause much movement of the fronts. Typical-

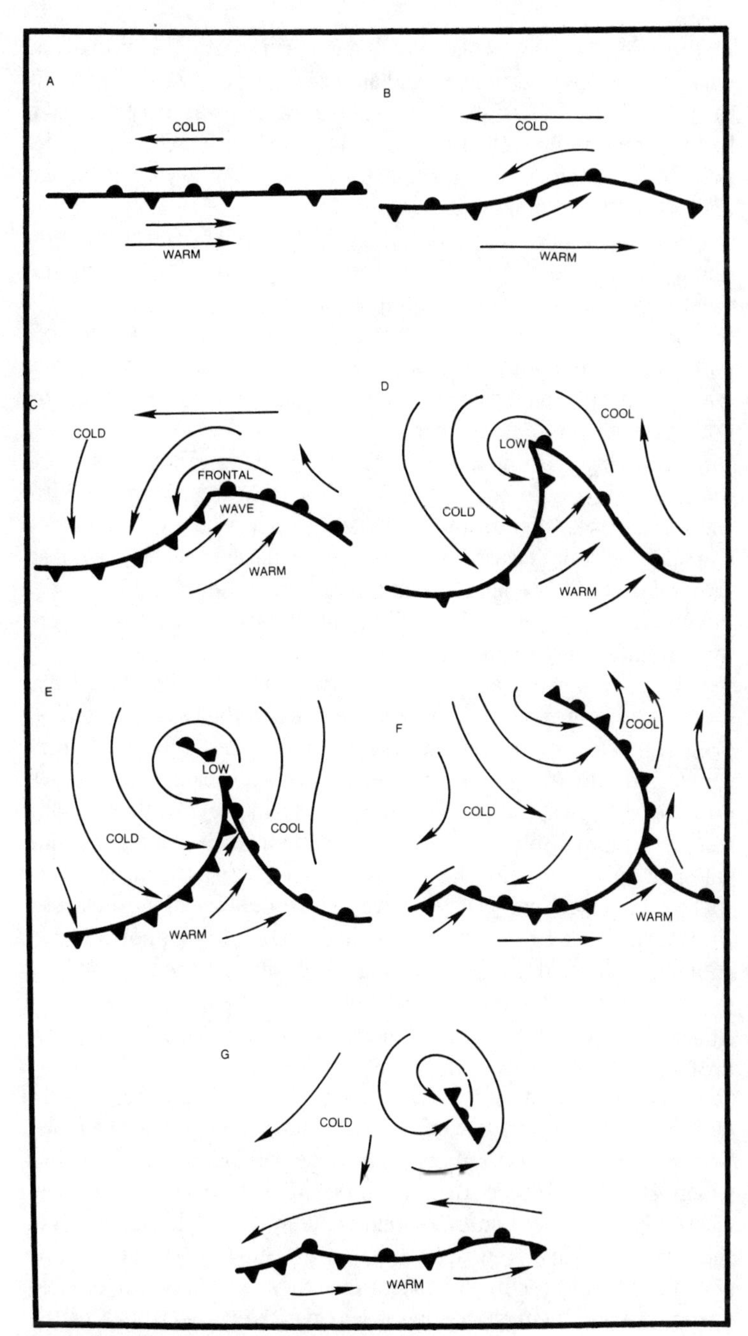

Fig. 5-2. Life cycle of a frontal wave.

ly, the cold front moves faster than the warm front. Finally, the cold front catches the warm front and an occluded front, or occlusion, is formed. This is very significant because when the storm occludes it has reached its maximum intensity as shown in E. This is not the best time to venture a trip through the fronts, as winds are strongest and precipitation most abundant.

As the occlusion grows in length, the cyclonic circulation diminishes in intensity and the fronts begin to form along the westward-trailing portion of the cold front.

In the final stage the two fronts may slow down and become a stationary front, as at G. The occlusion remnant is disappearing. The stage is set to begin again unless pressures continue to equalize on both sides of the stationary front.

How does all this affect a weather briefing? Well, it is more important, in the author's opinion, to know the big picture and the trends than it is to know the terminal forecast for every point along the route of flight. This is not to say that terminal forecasts are not an integral part of a preflight briefing. They are. They should be used, however, to bolster the faith in the big picture forecast or to spot local weather disturbances.

In contemplating a return trip from cleveland to Traverse City, Michigan, I switched on the television to catch the evening weather program. The meteorologist was describing a strong cold front running through the Cleveland area and was forecasting it to move on by early morning. The weather was to improve rather quickly from the moderate to heavy snows we had been getting. Having watched the weather for the last several days, it was apparent to me that the cold front was just reaching its mature stage in the life cycle. The warm front to the north and east in Ontario had not yet succumbed to the cold front. They would probably occlude the following day, but that would be east of Cleveland and we would be spared.

The next morning I tuned in to the "Today Show." The prognostic chart that they display is the forecast of positions of fronts at about noon Eastern Time. Sure enough, they had the front southeast of Cleveland and occluding at a point north of Niagara Falls. There were no successive fronts close behind on the weather prognostic chart, so the trip to northern Michigan looked clear. The call to Flight Service confirmed that conditions were ideal, except in the Traverse City area where they were having scattered snow showers as indicated by the hourly sequence. The terminal forecast also called for them to continue. Having a very good idea of the general position of all fronts in the Midwest, I was puzzled—at least to the point of looking for another reason for the precipitation. Had the terminal forecasts not been checked, a gap in flight information would have arisen. Traverse City sits on two large bays and is very near Lake Michigan. The snow shower activity in the area was due to lake effect—a local condition.

This anecdote should demonstrate the value of keeping up with trends in the weather along with checking all the other available reports. The pilot who doesn't take weather seriously and study it conscientiously before each trip is leaving a lot to chance.

Many of us, though, may not have time to watch television before a trip. Businessmen-pilots, for example, find that they must make a trip on the spur of the moment in order to service a long-distance customer.

What does a pilot need to ask for to get the most out of his weather briefing?

A phone call to Flight Service is the way most of us get our weather briefing. Few are the times when a National Weather Service Office or a Flight Service Station is close at hand. As a result we are kept from gazing at the weather depiction chart (the chart that shows the weather IFR or VFR as it is) at our leisure. We also cannot see the surface prognostic chart which predicts the movement of fronts for each 12-hour period. If we are aware that those charts are there next to the briefer, we can ask for that information from them.

The best way to kick off a good weather briefing is to first give the airplane type and identification number. Then specify whether you and your plane are equipped to fly IFR if necessary. Following those few words, describe the intended route of flight and request that the briefer begin with the area forecast. Many briefers, in fact most, don't like to give the area forecast. It involves extra effort on their part and many seem disgruntled when I ask. However, it is our taxes that pay their salaries.

The area forecast is very important, especially when a pilot may not have heard or seen a weather report in days. The area forecast helps the business-man-pilot shift gears, so to speak, from executive to professional pilot. The big picture for the geographic area can be gleaned quickly from the FA (area forecast). It usually outlines the positions of lows, highs and frontal systems. It gives the expected ceiling and cloud bases across the area as well as areas expected to receive precipitation. The freezing level is always described; in winter flying probably no other information is as important. The FA will also sketch the areas and altitudes where icing can be expected.

For a flight in cold weather, the area forecast really helps.

Next, you can let the briefer spiel off his normal routine. They like that. In his normal jag, you can expect to get, without asking, all hourly sequence reports and terminal forecasts for major terminals between your departure point and destination. Sometimes these guys will also throw in the winds aloft. If they don't, ask for them anyway.

From this point any other information will come on a strictly "ask for" basis. If you have gotten the position of the nearest front from the area forecast, then that information need not be requested again. However, it is necessary to ask what the expected movement of the front will be. The briefer will get that information from one of two different places: the surface prognostication chart or sometimes from a terminal forecast that predicts frontal passages. With this information a pilot can, in his own mind, make a forecast of where weather should improve or deteriorate.

Next, a pilot should request information on any Airmets or Sigmets. These outline the areas of low visibilities and ceilings as well as icing forecast. If a Sigmet is issued for icing on your route of flight, my advice is to forget the flight. Forecasts of severe icing are usually accurate for some reason. Many times I have seen North Central Airlines cancel for severe icing. When the airlines are not flying because of ice, no one should be: they cancel only for the worst imaginable conditions.

Generally, icing is forecast to be worse in clouds where precipitation is falling. For this reason, I usually leave the radar summary for last. The areas outlined for icing are fresh in one's mind and following it up with information from the radar summary will emphasize where the heaviest ice will be.

Another useful tip is to have handy a blank, pocket-to clipboard-sized, map of the United States. As the briefer describes the various weather features to you, they can be drawn on the map. The result is, you have the important areas highlighted on a map for ready reference in the cockpit.

Cold Fronts

Flying through a cold front is difficult in winter. The fact of the matter is, you shouldn't do it if you don't have to. Cold fronts have steep frontal slopes and the significance this holds for the aviator is that the weather can change abruptly. Approaching a cold front from the south side, the weather may be sunny and smooth. Once in the area of the cold front, however, the clouds tend to stack up quickly. The layers may go from near the surface to well above the tropopause. Lifting action is associated with all fronts, but a cold front produces the most instability. Thus, the heaviest turbulence and heaviest precipitation are associated with cold fronts.

As we mentioned earlier, in the area of the cold front, clouds stack up quickly. It has been found that moderate precipitation requires a cloud thickness of about 4,000 feet. This is notable and important to those who operate light aircraft. Transitioning through a cold front may be difficult because of ice in the clouds. If the ice is in moderate precipitation, then cloud thicknesses may preclude any possibility of climbing to a between-layers condition.

With moderate precipitation falling, careful monitoring of the outside air temperature is compulsory. When precip is moderate, ice buildup on the airframe can be severe. Therefore, choice of altitudes can be critical.

We know that cold fronts lift warm air aloft. We also know that the frontal slope is steep. Therefore, to climb and stay in warmer air in a cold front would be futile in a lightplane. In order to escape icing in a cold frontal situation, one must weigh the facts. If ice is already accumulating on the airframe, a climb may be out of the question. This may be due either to an already high operating altitude where little excess horsepower is left for climbing or to the destruction of lift by the ice already present. In such a situation a descent should be begun as soon as possible. Hopefully either an ice-free altitude will be found below or the descent can be continued to a precautionary approach and a landing made at the nearest airport.

Under some icing conditions a climb is possible. In cold front conditions, don't expect warmer air above. Expect colder air—so cold that freezing no longer occurs. An outside temperature below 15° F. at least, will be necessary. Also, the airframe will need to be chilled sufficiently so that frost does not form.

On the positive side of things, cold front activity often is not very wide. All that means is that the problems of icing may not last the entire trip. This is not to say that a pilot should continue onward foolishly in the hope that the sky will soon be blue ahead.

The easiest way for a pilot to determine if he has passed through the cold front is to watch the temperature gauge; a significant temperature drop will be noticed once through the front. Also, a wind correction to the right will be needed. This applies to any front no matter what the direction of flight. In the northern hemisphere, a right-hand wind correction is needed after frontal passage.

Warm Fronts in the Wintertime

Almost any one can recognize a cold front in the winter. The blustery winds and snows that have the audacity to cover our cars and homes are all part of it. However, recognition of the warm front may be one of the finer arts of piloting an aircraft cross-country.

No two warm fronts are the same. The combination of dangers of structural icing and low ceilings and visibilities almost never present themselves in a manner that makes a "go or no-go" decision easy. For that matter, a 180° turn usually is not an easy decision at which to arrive. As a result, the cold weather aviator must arm himself with some basic tools that will enable him to soar through the ragged wisps of clouds and decide whether to continue or land.

As compared to cold fronts, warm fronts have shallow frontal slopes. They are caused by warmer air overtaking and riding up over the trailing edge of cooler air. The shallow slope causes widespread cloudiness and low visibilities, often for hundreds of miles.

Cirrus clouds are the vanguard of the warm front. They appear where the warm air has been forced aloft up the slope of the cold air to an altitude of 20,000 feet or more. The average slope of a warm front is one percent or less. Thus, the first high clouds may be found as far as 500 miles or more in advance of the front. Approaching a warm front from the north side, a pilot will see clouds in a predetermined sequence. If we are familiar with this sequence, we can predict the changes that follow. The first clouds encountered are cirrus, followed by cirrostratus, altostratus and nimbostratus in that order. Precipitation may begin about 300 miles from the front.

The warm sector, which we mentioned earlier, is the wedge of air to the south of the warm front. The wind changes from an east or southeasterly direction on the north side of the front, to a southwesterly direction. The temperature will rise and, again, we will need a wind correction to the right after passing the front. If the warm sector is moist and the air unstable, there are likely to be snow showers if surface temperatures permit. Low ceilings and visibilities may persist after frontal passage in an inversion type effect with low cloud tops.

There are two sides to every frontal system. When considering a warm front, a pilot must be alert. From the south side, a warm front looks very much like a cold front and a VFR pilot would not Likely be enticed to penetrate the system. On the north side, however, a VFR pilot flying from good weather may be tempted to continue because the first signs look harmless enough. In the winter, however, warm fronts are freezing rain or drizzle producers. The rain will usually start coming down miles before the front where the pilot is flying under an umbrella of altostratus or altocumulus. Depending on the temperature aloft, icing may become a problem from the freezing drizzle. If temperatures are quite cold, then light snow will substitute for the freezing drizzle. The visibilities will then become low, giving the VFR pilot something to think about.

The IFR pilot may have it made for a while if he is traveling toward the front from the north. But as the frontal slope gets nearer the earth there is less time for the warm precipitation to be changed to a solid. Then, until the front passes, it becomes rain even though temperatures remain low at the lower altitudes. What the IFR pilot must do is climb to get on top of the frontal slope where all the air is warm and above freezing. This is usually, but not always, possible.

Warm fronts produce predominantly rime ice due to the stratiform clouds. As the pilot flies toward the front, though, cumuliform clouds are encountered which can cause severe clear icing. If the temperature at the surface is as cold as 14° F. and the warm front is close at hand, icing may be expected from within 300 to 400 feet of the surface to above 15,000 feet in the cumuliform clouds. This leaves a large part of the General Aviation fleet on the ground.

A low pressure center during the winter months is a blizzard builder. The warm moist air is pumped up over the top of the cold air and around to the northeast side. The moist air reaches its dewpoint and begins to fall into the cold air below, resulting in snow at the lower altitudes. Beware the northeast side of a low in the winter; this area is frequently the best for heavy snow accumulations. As the front occludes, the storm gains its greatest fury. On these days, it is best not to go if at all possible.

Shedding the Ice

The hard thing about learning to fly in icing conditions is that there are no rules of thumb. Some old pros always say to climb, but in light aircraft, a climb is not always possible. So, there goes that rule. If there is a secret to flying ice, it is the combination of a good preflight briefing and experience. No one is born with experience, so at some point in time we are forced to get our feet wet.

The key to knowing where to go to get out of ice can be found in the winds aloft forecast. Along with wind direction and velocities, temperatures are estimated. It is normal for the temperatures to decrease with an increase in altitude; but, unfortunately for those of us who fly in the slop a lot, that isn't always the case. Many times inversions form and warmer temperatures can be found. Temperature inversions can fall anywhere in the atmosphere. They are most likely to be found in the winter, however, between 4,000 and 9,000 feet.

The question is simple—whenever an inversion is predicted, what do we do about ice? Of course, the temperature above the inversion must be above freezing, or right at it, to do us any good. Even when a pilot has anti-icing and de-icing equipment, he should endeavor to find an altitude free of ice.

Often ice is associated with flying in clouds. We spoke of rime ice earlier. Rime ice is a fine granular ice that appears milky-white in color. It usually builds forward into the slip stream. Knowing what kind of ice is attacking the airframe can be useful in determining how to avoid it. For example, rime ice is associated with stratiform clouds. Stratiform clouds are typically layered and not excessively thick under most circumstances. A haven from ice, then, can sometimes be found between those stratus layers.

Let's say you are cruising at an altitude of 5,000 feet. At this altitude, the clouds are solid and hard-core instrument flying is required. On the climb up to 5,000 feet you topped one layer and climbed into the one you are operating in now. The temperature aloft forecast shows an inversion above freezing in the area of 6,000 feet. What is the best way out of the ice?

In a situation like this a climb would not be foolhardy. Seven thousand feet will put you above the inversion in warmer air, probably, where there will be no ice. Also, since the top of inversions are frequently the tops of cloud decks, any precipitation encountered at 7,000 feet would not stick to the plane. Another point in favor of climbing first is knowing we can fly between layers below 5,000 feet—the proverbial "ace in the hole."

On days when no inversion above is expected, the same choice can be made. If there are no pilot reports of tops in the area, I usually try to find them. If the total cloud decks can topped, we can fly fat, dumb and happy in the sun. During climbouts, always remember where the layer tops were since some days it is the only place to go.

Clear ice is not so easy to handle. It is usually formed by precipitation or flight through cumuliform clouds. It is the nature of clear ice to have a high rate of accretion, often near the tops of these clouds. Because of this it is imperative to arrive at a decision quickly. A short climb, a change of 1,000 feet in altitude, may free the plane of the condition. Also, it is often clear above the tops of this type of clouds. If you are near the top of the deck and clear conditions exist a short distance above, the light will be intense from the sun. A climb should be initiated at once and the condition will vanish.

On the other hand, if the clear ice is due to precipitation you'll know it. The sleet and rain slamming against the windshield are warning enough. This type of icing usually turns out to be the most severe. Therefore, avoidance precautions should be taken at once. Many times the best place to go is down—down to the nearest airport for a cup of coffee and a call to your wife.

If a plane is equipped with anti-ice and de-ice equiment, the battle will go more smoothly because a pilot can spend more time looking for the right altitude. It is easy to see that de-icing equipment can take a great deal of stress off the pilot. It is well worth the money.

What some pilots may not know is there is a technique for using de-icing equipment to maximum advantage.

First, if the plane starts picking up ice, begin action immediately in selecting another altitude. All anti-ice should be turned "on" if it wasn't prior to entering the cloud. Among anti-ice are the hot props, pitot heat and electric windshield. This equipment will do much to prolong the flight at the initial cruising altitude if, for some reason, a change cannot be initiated right away due to terrain or conflicting traffic.

Secondly, let the ice build on the airfoils until 3/4-inch to one-inch is present on the leading edge. If expansion boots are used too early, the end result may be like not having boots at all. If the ice is not of sufficient thickness, the boots will not be able to pop it off. Sometimes, if the boots are exercised prematurely, they will expand the ice outward leaving a hollow space underneath. The ice will then continue to accumulate but will not be removed by further expansions because each expansion is the same size.

Once the ice has been broken off, let the ice accumulation begin again. Of course, continue efforts to find an ice-free altitude. Always be careful to allow about an inch to accumulate before exercising the boots.

Some expansion boot systems have an automatic cycle function. Unless ice accretion is occurring at an alarming rate, always operate the boots manually. Automatic expansion every 20 to 30 seconds is too much and may result in one of the conditions of non-performance mentioned above.

If you are new to flying ice or new at flying a twin-engine aircraft in ice, there are some sounds you can expect to hear that can be alarming. The most common noise is that of ice slinging off the props. The loud bang can be heard even when wearing a headset. Passengers should be warned that they might hear this sound.

Fig. 5-3. When the flying is over, always install the control lock to guard against winter's blustery winds.

Another sound comes from the long wire ADF antenna that runs from the top of the cabin to the vertical stabilizer. Once this wire accumulates ice it will begin to vibrate with a low pitch. The low frequency flutter is more of a viration than a sound, but you'll know what it is when you hear it.

Ice accumulation on a twin is like a visit to a house of horrors. The groans and creaks of the airplane are unusual and they can be very disturbing. Regardless of the type of aircraft, icing should be considered dangerous. The best time to do something about ice is right away. With immediate action, navigation through icy areas can be achieved with regularity and safety. The problems of icing in the winter can be handled much easier than those of the thunderstorms of summer.

Chapter 6
Survival

In the summer a crash landing may not have the implications that the same crash would have in the winter. An emergency landing in a field in the summer might be followed by flagging down a farmer and riding to town. In the winter, a person might have to spend a couple of days and nights with the plane, unable to walk out in deep snow. It is for this reason that we will examine what can be done to maximize our survival should a crash landing occur in the winter.

After a crash landing, it is best to get away from the aircraft as soon as possible. Fire is very much a possibility due to ruptured gas tanks and hot engine parts. After moving away from the airplane, take time to analyze the situation and then begin to help others. Those who have injuries should be attended to first. Remember to stay away from the plane until all gasoline fumes have dissipated.

Next, sit down and think. Thoroughly analyze the problems and keep in mind that survival is 80 percent mental, 10 percent equipment and 10 percent skill. Since mental factors present the number one problem, a goal should be chosen to conquer regardless of the consequences. "Give-up-itis" or a do-nothing attitude will undermine the best chances of surviving. Put any ideas like these out of your mind. Although you have chosen a goal, take time to think through each problem (Table 6-1).

The mind can be an ally or it can be an enemy. In alien situations, people tend to imagine things that are not there. Every one of us has basic fears, but if we just take time to realize they are just

Table 6-1. Length of Time You Can Live Without Necessities.

Requirements for life:	You can live without it approximately:
Air	3 minutes
Body Shelter	6 hours in severe weather
Water	3-6 days
Food	3 weeks
Will to live	?

that—fear—we can deal with ourselves more rationally. These normal fears that one might experience in a survival situation are: fear of the unknown, fear of darkness, fear of discomfort, fear of being alone, fear of animals, fear of death, fear of punishment, and fear of personal guilt.

If your mind is the best tool for survival, you should also know that the biggest enemy is yourself. The number two enemy is injuries; number three is temperature; number four is disease. There are things that can be done to alleviate each of these problems, however.

The decision of whether to remain with the airplane or start out on foot may be extremely important. Before deciding, stop and think. Did you file a flight plan? If so, it may be better to let them search and find you. Is the emergency locator transmitter operating? Do you have a survival kit? Depending on the answers to these questions, a sensible decision can be made to stay or to go.

Whatever the decision, don't fight a storm. It is always best to stay put and find shelter during storms, though most rarely last longer than 24 hours.

A thorough evaluation should be made of items in the aircraft that can be used to aid in survival. If you have decided to start trucking out through the woods, rip the compass off the flight panel. It will keep you going in one direction. Any gasoline that may be left will help make a fire. A major cause of accidents is fuel starvation, however, so it may happen that you used all of it to get you this far. Oil can be used for smoke signals but this item may or may not be accessible. Seat upholstery makes good extra covering for the feet or hands. Wiring can be pulled out to use for tie strings to rig up a shelter. Finally, the battery can be used to ignite fuel if you happen to be a non-smoker and don't carry matches.

Another high priority item is body heat. Body heat is wasted from eating snow. It is much better to make a fire and heat water for drinking. This achieves two purposes; sterilization and conservation of body heat. It is not impossible to conserve enough energy to

Table 6-2. Wind Chill Chart.

	ACTUAL THERMOMETER READING °F										
Estimated Wind Speed mph	*50*	*40*	*30*	*20*	*10*	*0*	−10	−20	−30	−40	−50
	EQUIVALENT TEMPERATURE °F.										
Calm	50	40	30	20	10	0	−10	−20	−30	−40	−50
5	48	37	27	16	6	−5	−15	−26	−36	−47	−57
10	40	28	16	4	−9	−21	−33	−46	−58	−70	−83
15	36	22	9	−5	−18	−36	−45	−58	−72	−85	−99
20	32	18	4	−10	−25	−39	−53	−67	−82	−96	−110
25	30	16	0	−15	−29	−44	−59	−74	−88	−104	−118
30	28	13	−2	−18	−33	−48	−63	−79	−94	−109	−125
35	27	11	−4	−20	−35	−49	−67	−83	−98	−113	−129
40	26	10	−6	−21	−37	−53	−69	−85	−100	−116	−132
Wind speeds greater than 40 mph have little additional effect	LITTLE DANGER FOR PROPERLY CLOTHED PERSON						INCREASING DANGER			GREAT DANGER	
							DANGER FROM FREEZING OF EXPOSED FLESH				

Fig. 6-1. The Big Red Bag shown with the "Grab Bag" from Delta Air Lines and the AOPA Survival Kit, for comparison. Which would you want if you went down in the wintertime?

survive three weeks if you have water and stay dry. Clothing loses most of its insulative value when it becomes wet and body heat can escape 240 times faster in wet clothing than in dry clothing. To replace energy, it is best to eat small amounts of sugary foods (Table 6-2).

Survival Kits

By now, you probably have thought far enough ahead to survival kits. To be found in a situation needing one could happen to any of us. The weight penalty in the airplane is negligible, yet the value in peace of mind is priceless (Figs.6-1 through 6-3).

Listed below are some things that would make a simple survival kit. Most or all of these items can be found in the home or the garage. A more complex type of kit will be described later.

1. First you need a metal container with a lid. This container can be used to heat water, make tea, use as a digging tool or (polished) as a signal mirror.

2. Boy Scout knife.
3. Small candle.
4. Box of matches (wrapped in plastic).
5. Leaf bag (pull over head, cut hole for face).
6. Garbage bag (step in, pull up and tuck in pants or tie around waist). Be sure to cut holes for your legs. Both bags together will protect the body from heat loss.
7. Sugar cubes (wrapped in plastic, 6 to 12 ought to do it).
8. Plastic tape.

This is only a sample of what can be done with a few household goods. The following is a more expansive type of kit. These items can be bought commercially or you can collect them yourself.

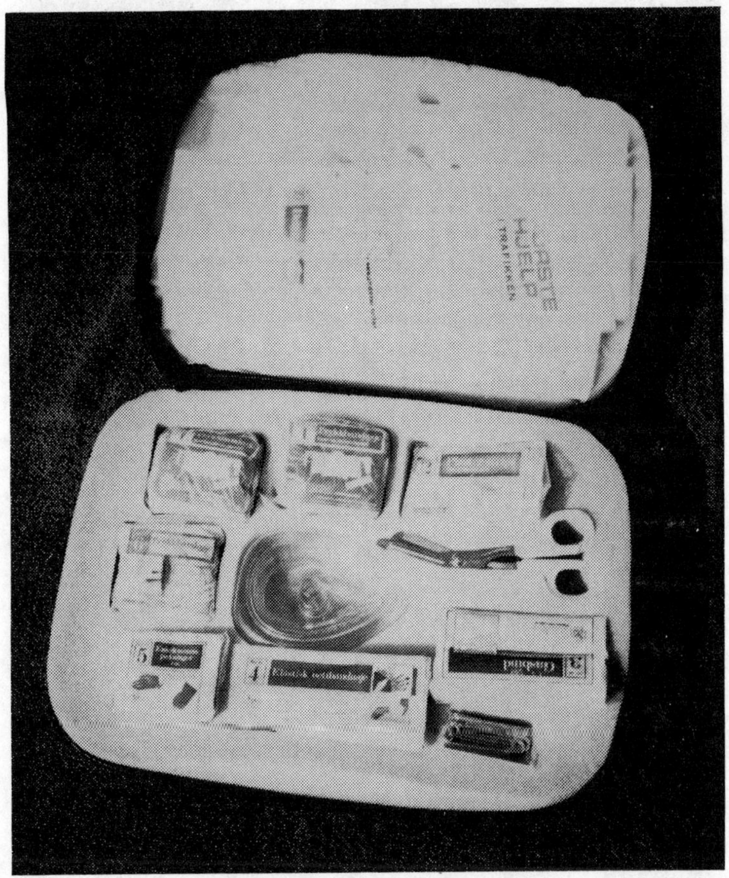

Fig. 6-2. A first aid kit shaped like a small briefcase. The equipment is encased in foam rubber and the kit can be used as a pillow or cushion.

Fig. 6-3. Our "Grab Bag," or basic survival kit, with the hand axe, machette and the take-down 12-gauge shotgun in the foreground. With the equipment pictured here and our normal supply of water, we can survive for a week in a mild climate.

Container: Any lightweight metal container with lid, suitable for heating and storing water.

Life Support Tools:

Hack saw—Single handle with wood blade and metal blade.
Pliers—both the vise grip and slip-joint types.
Several screwdrivers.

Shelters (minimum of 2)

Large plastic sheets (9′ × 12′)—should be of heavy gauge and colored red, yellow or orange for signal panels.

First Aid Kit

Sealable Plastic Container

2 Compress bandages, 1 Triangle bandage, Small Roll 2″ tape, 6 3″ × 3″ gauze pads, 25 Aspirin, 10 Bandaids, Razor blades or scissors, hotel-size soap, Kotex (purse size), Kleenex (purse size), 6 safety pins, 1 small tube of Unguentine or Foile.

Life Support Kit

Water proofed matches, candle or fire starter, signal mirror, compass, knife (Boy Scout style), insect repellent, mosquito net, 50′ of 1/8″ nylon rope or shroud line, whistle, smoke flares or red day-nite flares.

Fig. 6-4. When the lean-to is rigged with a tarpaulin and closed at the ends, it offers snug quarters.

Fig. 6-5. A quickly detachable emergency locator transmitter (ELT) with a plug-in microphone will make direct radio communications with rescue aircraft possible. A small battery-powered VHF receiver will make two-way communications possible.

Food and Energy Package—1 man 5-day rations

2 or 3 cans of Sego nutriment or Metrecal for liquid and energy, 30 sugar cubes (wrapped), 10 pilot bread or 25 crackers, 10 packets of salt, 3 tea bags, 12 rock candy, 5 gum, 10 bouillon cubes and 20 protein wafers.

*Use poly bags for water storage. Put each item in small plastic bags and seal. Put everything in small metal can (cook pot) and seal with poly bag and tape.

These lists can be added to or modified as you wish. Hopefully, we will never need to take these items out of their containers, but being prepared is worth a lot of peace of mind (Figs. 6-4 and 6-5).

Good flying this winter!

Glossary

Glossary

abort—To terminate a preplanned aircraft maneuver; e.g., an aborted takeoff.

acknowledge—Let me know that you have received and understand my message.

additional services—Advisory information provided by ATC which includes, but is not limited to, the following:

1. Traffic advisories;
2. Vectors, when requested by the pilot, to assist aircraft receiving traffic advisories to avoid observed;
3. Altitude deviation information of 300 feet or more from an assigned altitude as observed on a verified (reading correctly) automatic altitude readout (MODE C);
4. Advisories that traffic is no longer a factor;
5. Weather and chaff information;
6. Weather assistance;
7. Bird activity information;
8. Holding pattern surveillance.

Additional services are provided to the extent possible contingent only upon the controller's capability to fit them into the performance of higher priority duties and on the basis of limitations of the radar, volume of traffic, frequency congestion and controller workload. The controller has complete discretion for determining if he is able to provide or continue to provide a service in a particular case. The controller's reason not to provide or continue to provide a service in a particular instance is not subject to question by the pilot and need not be made known to him.

(I'd like to add that in non-mandatory radar service areas where a pilot feels he is getting unnecessary vectors, he can cancel ATC service as well. In this case the pilot need not tell the controller the reason. As you can see, it is a two-way stret.)

advisory service—Advice and information provided by a facility to assist pilots in the safe conduct of flight and aircraft movement. These service areas are not necessarily government-run.

aeronautical beacon—A visual navaid displaying flashes of white and/or colored light to indicate the location of an airport, a heliport, a landmark, a certain point of a Federal Airway in mountainous terrain, or a hazard.

air carrier district office—An FAA field office serving an assigned geographical area, staffed with Flight Standards personnel serving the aviation industry and the general public on matters related to the certification and operation of scheduled air carrier and other large aircraft operations.

aircraft classes—For the purposes of Wake Turbulence Separation Minima, ATC classifies aircraft as follows:

1. *Heavy*—Aircraft capable of takeoff weights of 300,000 pounds or more, whether or not they are operating at this weight during a particular phase of flight.
2. *Large*—Aircraft of from 12,500 pounds, maximum certificated takeoff weight, up to 300,000 pounds.
3. *Small*—Aircraft of 12,500 pounds or less, maximum certificated takeoff weight.

airmet/airman's meteorological information—Inflight weather advisories which cover moderate icing, moderate turbulence, sustained winds of 30 knots or more within 2,000 feet of the surface and the initial onset of phenomena producing extensive areas of visibilities below three miles or ceilings less than 1,000 feet. It concerns weather phenomena which are of operational interest to all aircraft and potentially hazardous to aircraft having limited capability because of lack of equipment, instrumentation or pilot qualifications.

Automated Radar Terminal Systems/ARTS—The generic term for the ultimate in functional capability afforded by several automation systems. Each differs in functional capabilities and equipment. ARTS plus a suffix Roman numeral denotes a specific system. A following letter indicates a major modification to that system. In general, an ARTS displays for the terminal controller aircraft identification, flight plan data, other flight associated information such as altitude and speed, and aircraft position symbols in conjunction with his radar presentation. Normal radar co-exists with the alphanumeric display. In addition to enhancing visualization of the air traffic situation, ARTS facilitate intra/inter-facility transfer and coordination of flight information. These capabilities are enabled by specially designed computers and subsystems tailored to the radar and communications equipments and operational requirements of each automated facility. Modular design permits adoption of improvements in computer software and electronic technologies as they become available while retaining the characteristics unique to each system.

1. *ARTS IA*—The functional capabilities and equipment of the New York Common IFR Room Terminal Automation System, it tracks primary as well as secondary targets derived from two radar sources. The aircraft targets are displayed on a radar-type console by means of an alphanumeric generator. Aircraft identity is depicted in association with the appropriate aircraft target. When the aircraft is equipped with an encoded altimeter, its altitude is also displayed. The system can exchange flight plan information with the ARTCC.
2. *ARTS II*—Programmable non-tracking computer aided display subsystems capable of modular expansion, ARTS II systems provide a level of automated air traffic control capability at terminals having low to medium activity. Flight identification and altitude may be associated with the display of secondary radar targets. Also, flight plan information may be exchanged between the terminal and ARTCC.
3. *ARTS III*—The Beacon Tracking Level (BTL) of the modular programmable automated radar terminal system in use at medium- to high-activity terminals, ARTS III detects, tracks and predicts secondary radar derived aircraft targets. These are displayed by means of computer generated symbols and alphanumeric characters depicting flight identification altitude, ground speed and flight plan data. Although it does not track primary targets, they are displayed coincident with the secondary radar as well as the

symbols and alphanumerics. The system has the capability of communicating with ARTCC's and other ARTS III facilities. ARTS III is found at all Group II and III TCA's.

4. *ARTS IIIA*—The Radar Tracking and Beacon Tracking Level of the modular programmable automated radar terminal system, ARTS IIIA detects, tracks and predicts primary as well as secondary radar derived aircraft targets. An enhancement of the ARTS III, this more sophisticated computer driven system will eventually replace the ARTS IA system and upgrade about half of the existing ARTS III systems. The enhanced system will provide improved tracking, continuous data recording and fail-safe capabilities.

displaced threshold—A threshold that is located at a point on the runway other than the designated beginning of the runway.

Emergency Locator Transmitter/ELT—A radio transmitter attached to the aircraft structure which operates from its own power source on 121.5 MHz and 243.0 MHz. It aids in locating downed aircraft by radiating a downward sweeping audio tone, two to four times per second. It is designed to function without human action after an accident.

feathered propeller—A propeller whose blades have been rotated so that the leading and trailing edges are nearly parallel with the aircraft flight path to stop or minimize drag and engine rotation. It is normally used to indicate shutdown of a reciprocating or turboprop engine due to malfunction.

final—Commonly used to mean that an aircraft is on the final approach course or is aligned with a landing area.

General Aviation District Office/GADO—An FAA field office serving a designated geographical area, staffed with Flights Standards personnel who have responsibility for serving the aviation industry and the general public on all matters relating to the certification and operation of general aviation aircraft.

minimum fuel—Indicating that an aircraft's fuel supply has reached a state where, upon reaching the destination, it can accept little or no delay, this is not an emergency situation but merely an indication that an emergency situation is possible should any undue delay occur.

Sigmet/significant meteorological information—A weather advisory issued concerning weather significant to the safety of all aircraft, SIGMET advisories cover tornadoes, lines of thunderstorms, embedded thunderstorms, large hail, severe and extreme turbulence, severe icing, and widespread dust or sandstorms that reduce visibility to less than three miles.

turbojet aircraft—An aircraft having a jet engine in which the energy of the jet operates a turbine which, in turn, operates the air compressor.

turboprop aircraft—An aircraft having a jet engine in which the energy of the jet operates a turbine which drives the propeller.

VFR not recommended—An advisory provided by a Flight Service Station to a pilot during a preflight or inflight briefing that flight under Visual Flight Rules is not recommended. To be given when the current and/or forecasted weather conditions are at or below VFR minimums, it does not abrogate the pilot's authority to make his own decision.

wind shear—A change in wind speed and/or wind direction in short distance resulting in a tearing or shearing effect, it can exist in horizontal or vertical or, occasionally, in both directions.

Appendix

Appendix
Three Letter Station Identifiers

Table A-1. Incomplete List of Airports In The United States Snow Belt.

AAA—Lincoln, Illinois
AAJ—Binghamton, New York
AAU—Ashland, Ohio
ABR—Aberdeen, South Dakota
ABQ—Albuquerque, New Mexico
ACB—Bellaire, Michigan
ACQ—Waseca, Minnesota
AEL—Albert Lea, Minnesota
AHT—Amchitka, Alaska
ALB—Albany, New York
ALO—Waterloo, Iowa
AMH—Nashua, New Hampshire
ASE—Aspen, Colorado
AUW—Wausau, Wisconsin
AXN—Alexandria, Minnesota
AZO—Kalamazoo, Michigan
BFA—Boyne Falls, Michigan
BIL—Billings, Montana
BIS—Bismarck, North Dakota
BJI—Bemidji, Minnesota
BKE—Baker, Oregon
BMC—Brigham City, Utah
BNO—Burns, Oregon
BOI—Boise, Idaho
BOS—Boston, Massachussetts
BRD—Brainerd, Minnesota
BRL—Burlington, Iowa
BRN—Mountain Home, Idaho
BTM—Butte, Montana
BYI—Burley, Idaho
BZN—Bozeman, Montana
CAD—Cadillac, Michigan
CAR—Caribou, Maine
CBP—Columbus, Ohio
CDR—Chadron, Nebraska
CLE—Cleveland, Ohio
CIM—Cimarron, New Mexico
COE—Coeur D'Alene, Idaho
CON—Concord, New Hampshire
COS—Colorado Springs, Colorado
CPR—Casper, Wyoming
CTB—Cut Bank, Montana
CVG—Cincinnati, Ohio
CZI—Crazy Woman, Wyoming
DDC—Dodge City, Kansas
DEC—Decatur, Illinois
DEN—Denver, Colorado
DET—Detroit, Michigan
DFI—Defiance, Ohio
DLH—Duluth, Minnesota

Table A-1. Incomplete List of Airports In The United States Snow Belt (continued from page 155).

DLS—The Dalles, Oregon
DRO—Durango, Colorado
DXR—Danbury, Connecticut
EAU—Eau Claire, Wisconsin
EKO—Elko, Nevada
ELM—Elmira, New York
ELN—Ellensburg, Washington
ESC—Escanaba, Michigan
EUG—Eugene, Oregon
EWB—New Bedford, Massachussetts
EWR—Newark, New Jersey
FDY—Findlay, Ohio
FFU—Provo, Utah
FGT—Farmington, Minnesota
FMN—Farmington, New Mexico
FNT—Flint, Michigan
FOD—Fort Dodge, Iowa
FSD—Sioux Falls, South Dakota
FWA—Fort Wayne, Indiana
GCC—Gillette, Wyoming
GEG—Spokane, Washington
GFA—Great Falls, Montana
GFK—Great Forks, North Dakota
GFL—Glens Falls, New York
GJT—Grand Junction, Colorado
GLR—Gaylord, Michigan
GNB—Granby, Colorado
GRB—Green Bay, Wisconsin
GRI—Grand Island, Nebraska
GRR—Grand Rapids, Michigan
GSH—Goshen, Indiana
GWS—Glenwood Springs, Colorado
HAR—Harrisburg, Pennsylvania
HFD—Hartford, Connecticut
HLN—Helena, Montana
HNN—Henderson, West Virginia
HTS—Huntington, West Virginia
HVR—Havre, Montana
HUL—Houlton, Maine
IAG—Niagara Falls, New York
IND—Indianapolis, Indiana
INL—International Falls, Minnesota
LAR—Laramie, Wyoming
MCE—Merced, California
MFR—Medford, Oregon
ORD—Chicago, Illinois

Index